Because Everyone

Deserves a Cure

6/2/19

A Cure Within

Scientists Unleashing the Immune
System to Kill Cancer

OTHER TITLES FROM COLD SPRING HARBOR LABORATORY PRESS

*Cold Spring Harbor Symposia on Quantitative Biology LXXVIII:
Immunity and Tolerance*

Orphan: The Quest to Save Children with Rare Genetic Disorders

*The Strongest Boy in the World: How Genetic Information Is Reshaping
Our Lives,* Updated and Expanded Edition

*Is It in Your Genes? The Influence of Genes on Common Disorders and Diseases
That Affect You and Your Family*

Abraham Lincoln's DNA and Other Adventures in Genetics

A Cure Within

Scientists Unleashing the Immune System to Kill Cancer

NEIL CANAVAN

The Trout Group LLC

COLD SPRING HARBOR LABORATORY PRESS

Cold Spring Harbor, New York • www.cshlpress.org

A Cure Within
Scientists Unleashing the Immune System to Kill Cancer

© 2018 by Neil Canavan; published by Cold Spring Harbor Laboratory Press
Printed in the United States of America

Publisher and Acquisition Editor	John Inglis
Director of Editorial Services	Jan Argentine
Project Manager	Inez Sialiano
Permissions Coordinator	Carol Brown
Director of Publication Services	Linda Sussman
Production Editor	Kathleen Bubbeo
Production Manager	Denise Weiss
Cover Designer	Mike Albano

Front cover artwork: Watercolor illustration of typical lymphocyte. (Illustration, Judy Cuddihy.)

Library of Congress Cataloging-in-Publication Data

Names: Canavan, Neil, author.
Title: A cure within : scientists unleashing the immune system to kill cancer / Neil Canavan, The Trout Group, LLC.
Description: Cold Spring Harbor, New York : Cold Spring Harbor Laboratory Press, [2017] | Includes bibliographical references and index.
Identifiers: LCCN 2017040410 (print) | LCCN 2017037866 (ebook) | ISBN 9781621822172 (hardcover : alk. paper) | ISBN 9781621822189 (ePub3) | ISBN 9781621822196 (Kindle)
Subjects: LCSH: Cancer--Immunotherapy--Research--History. | Oncologists--Biography.
Classification: LCC RC271.I45 C32 2017 (ebook) | LCC RC271.I45 (print) |
DDC 362.19699/4--dc23
LC record available at https://lccn.loc.gov/2017037866

10

For a complete catalog of all Cold Spring Harbor Laboratory Press publications, visit our website at www.cshlpress.org.

I dedicate this book to my parents, Greg and Jean, and my two brothers, Chris and Mark:

Greg Canavan (1922–1971)
Hodgkin's lymphoma (1962), non–small cell lung cancer

Jean Canavan (1927–2015)
cancer of the pancreas

Chris Canavan (1956–)
diffuse large B-cell lymphoma

Mark Canavan (1962–)
diffuse large B-cell lymphoma

I am now eight years older than my father ever got to be.

I have had it with this goddamn disease.

Contents

Foreword

The writing is on the wall: cancer's long and terrible reign as "The Emperor of All Maladies" is soon coming to an end. The disease that kills more than eight million people a year worldwide has met a new foe capable of outsmarting and defeating it, and it's been right under our noses the entire time: our own immune system.

Though still in its early days, cancer immunotherapy—also called immuno-oncology—has been hailed as a revolutionary treatment approach that is disrupting the status quo in cancer care. Mobilizing the immune system to recognize and attack cancer has been a dream that has long eluded the field, and it is only recently that unprecedented recoveries of patients with very advanced disease have effectively demonstrated immunotherapy's power to save lives.

Although such complete responses are yet the exception to the rule, they give new hope to patients and fuel the passion of the scientists, clinicians, and drug manufacturers who are working to realize immunotherapy's full potential.

How did we get here? Why is the world only now hearing about immunotherapy? The recent flurry of headlines about clinical successes and FDA approvals seems to have come from nowhere. This is hardly the case.

Cancer immunotherapy's story stretches all the way back to the late 1890s, and it has taken decades of basic research and billions of dollars of investment to build the foundation upon which today's lifesaving treatments are based. This book offers a uniquely entertaining yet inspiring glimpse into the lives and minds of the academic and industry pioneers who forged this new field. It is a story of how an obscure and oft-derided field of cancer research —and the tenacious few scientists who refused to abandon it—came from behind to become the new "darling of oncology."

> *This book offers a uniquely entertaining yet inspiring glimpse into the lives and minds of the academic and industry pioneers who forged this new field.*

As the CEO of the Cancer Research Institute (CRI), the first and, up until recently, the only nonprofit organization dedicated exclusively to advancing immuno-oncology, I have had a front-row seat to the exciting developments as this field has grown from the earliest laboratory discoveries into today's headline-making breakthroughs. I also am very proud of the fact that nearly all of the individuals featured in this book are in some way connected to CRI, whether as members of our scientific leadership and/or as recipients of funding from our research programs. Many consider CRI's support a lifeline that enabled them to continue pursuing their important research.

Throughout this book, a name appears of another pioneer who not only contributed seminal discoveries to the field in his own right, but who also touched so many others, guiding them and supporting their research and encouraging them to press forward. That man was Lloyd J. Old, M.D., whose long and distinguished career as a scientist, mentor, and visionary leader earned him the title "Father of Modern Tumor Immunology." His many contributions to the field include the discovery of the first link between the major histocompatibility complex (MHC) and disease (leukemia); discovery of the role of Epstein–Barr virus (EBV) in the development of nasopharyngeal cancer; the discovery of tumor necrosis factor (TNF); and his definition of the concept of cell-surface differentiation antigens, a key to our understanding of how the immune system and its various cellular components function. These are but a few of his many notable accomplishments.

Lloyd was a wise mentor to me, as he was to so many others. But he was also a friend and confidant and was always a gentleman and scholar. As the director of the Cancer Research Institute Scientific Advisory Council, he played a pivotal role in guiding CRI's programs and future directions. When he assumed this role in 1970, his first action was to persuade the world's leading immunologists, including Nobel Prize winners and members of the National Academy of Sciences, to join CRI as advisors. Together, they established the primacy of basic immunology research as the first and most important step toward the eventual immunological control of cancer. By association with these scientific luminaries, CRI's reputation in the field rocketed, enabling us to attract the finest minds to this work.

In 1971, Lloyd established CRI's Postdoctoral Fellowship Program, as he devoutly believed in young scientists and the need to train a generation of immunologists. This program has supported more than 1300 scientists since then, many of whom have gone on to achieve prominence in the field and who have, in turn, trained generations of other young minds. More importantly, their work has produced fundamental knowledge that provides a scientific rationale for human trials of experimental immunotherapies.

Lloyd's prescience and vision would lead to other CRI programs, each designed to fill a critical need in the spectrum of scientific discovery and drug development. The one he was most proud of, however, was the Cancer Vaccine Collaborative. Through this initiative, he forged a partnership between CRI and the Ludwig Institute for Cancer Research, an organization he led as director and chief executive for nearly 20 years. This partnership enabled a global network of expert tumor immunologists to carry out coordinated, multicenter, single-variable, parallel, academic clinical trials that employed standardized correlative immunological monitoring designed to extract the most meaningful data from first-in-human studies. It was the first network of its kind in the cancer immunotherapy space.

Tragically, in 2011, Lloyd died from prostate cancer at the age of 78. It's a cruel irony not only that the disease he'd dedicated his life to defeating ultimately took him from us, but also that it should happen just as immunotherapy was beginning to prove itself and take its first steps toward the limelight.

No one person can be credited with the successes of an entire field. Lloyd would certainly never say so, as he often cast credit aside, but I think his indelible imprint is evident across the careers and lives of many in this field, whether directly or indirectly. I do not believe the field would be where it is today if it were not for his vision, leadership, and pursuit of scientific excellence.

It is gratifying that the field is now rich with other distinguished scientists whose own seminal work has helped us to come closer to immunologically based cures for all cancers. I trust that you will enjoy learning about them, their work, and this powerful new way to treat cancer as you explore the chapters ahead.

JILL O'DONNELL-TORMEY, PH.D.
CEO and Director of Scientific Affairs
Cancer Research Institute
New York, New York
August 2017

Acknowledgments

To paraphrase a recent political sentiment: This book—it is not mine. This is a We book if ever there was one.

First, and most obviously, I would like to thank the direct participants, the scientists/clinicians who dedicated their lives to revealing the dizzying kaleidoscope of molecular interactions that keep us all alive and who are now using those revelations to relieve human suffering.

The many years of training, the many hours of work each day, the weekends and holidays spent in the lab, the genius of insight, the tenacity of being, the resilience of spirit in the face of the deadly, the urgent, and the unknown—such realities are the résumé of these fine, fine people.

I do not have such stuff. I wish I did. To each and all I say: I stand in awe.

That said, I also stand somewhat defensive. There are many well-deserving investigators not celebrated in these pages, and to them, with hat in hand, I apologize. As I hope I have made clear in the book, there is no discovery—beyond that of finding your own navel—that originates from the realm of a single individual. Science is a continuum, an accumulation, a symphony; should a soloist arise in a spotlight they would soon resume their seat. So again, to the rest of the orchestra I say—I hear you, I see you, and I acknowledge that without you, there is no music.

Next, there are the countless many that shall remain anonymous: the patients. Without the patients willing to participate in clinical trials, there are no data; there were no breakthroughs; there are no cures. Were it not for those individuals who were brave enough to sign up for *human experiments*, we would still be treating syphilis with powdered mercury.

Returning to direct contributors to the book itself, I would like to thank Dr. Joshua Brody, currently of New York's Mount Sinai Hospital, whom I relied on to ensure that my words made good scientific sense. Josh, I owe you.

Others who contributed in a variety of important ways include Gabe Dolsten, Max Choulika, Brittany Correia, Courtney Powell, Maria Alexander, Mindon Laue, Stephen Rego, Patrick Rivers, Meryl Houghton, Sarah Weisbrod, Meggie Purcell, Rebecca John, and Michael Gibralter, and, at Cold Spring Harbor, John Inglis, Mala Mazzullo, Linda Sussman, Denise Weiss, Kathleen Bubbeo, Jan Argentine, Inez Sialiano, Wayne Manos, and Rob Redmond.

On a very personal note, I would like to thank Dani Joy Gero-Brett, whose middle name gives it all away. Without her love and boundless affections, I would have shot myself long ago.

Last, I would like to thank my boss, Jonathan Fassberg, founder of The Trout Group. Were it not for his foresight, insight, encouragement, and financial support this book would not exist, because—God knows—I did not want to write it.

Book writing is hard. Writing about cutting-edge science is hard. Writing a book about cutting-edge science by interpreting the spoken words of cutting-edge scientists, and then translating those words into a palatable read . . . well, suffice it to say that in the nearly two years it took me to make it all happen, I've slept a lot less and drank a good bit more.

Through it all, I had Jonathan's unwavering support. He saw the coming medical revolution, and he knew it was vital to get the word out—to investors, to clinicians, to the patients themselves—anyone and everyone. He wanted everyone to understand what was happening and to come to know the visionaries making it happen. He hired me, he trusted me, and he gave me every possible freedom to tell the story the way I thought it should be told.

The result is this: Jonathan has my eternal and profound gratitude, and the readers now have in their hands the most important words I have ever written.

Introduction

"It works where nothing worked before."
—Gordon Freeman, Ph.D., Dana-Farber Cancer Institute

The way we treat cancer is about to change forever. This revolution—and it is precisely that—was sparked not by the invention of a new drug but by the advent of an entirely new way of thinking about and managing cancer patients. Going forward, doctors will not use pharmaceuticals to attack a tumor, at least not directly. Rather, the oncologist will treat the patient's *immune system* with a drug, thereby enabling it to track down and destroy the cancer.

This new branch of medicine is called immuno-oncology (IO), and to date the results from using this approach to treat cancer are without precedent.

In brief, immuno-oncology is based on the idea that, just like with any bacterial or viral infection, the human immune system is capable of recognizing, attacking, and killing tumor cells. This realization is not exactly new; what is new is that the immune system's potential for ridding the body of cancer is now being deliberately employed.

A bit of history: In the early 1900s a renowned surgeon from New York City named William Coley read about the case of a cancer patient who came down with a near-fatal, postoperative infection. That event was routine enough, but what made the case provocative was that the patient not only survived the infection, but that shortly thereafter all his remaining inoperable tumors disappeared. Coley was particularly struck by this history because he had recently operated on a patient with a very similar cancer—a patient that came though surgery with flying colors without ensuing infection, only to later die of the residual cancer that the surgery failed to remove.

After seeking out and finding a number of cases where tumors spontaneously regressed after the patient experienced a bout of infection, Dr. Coley explored and expanded on these findings and went on to develop a related cancer treatment, a bacterial preparation later dubbed "Coley's Toxins."

Unfortunately, the Toxins were only marginally effective. No one had any idea how they actually worked—if they worked at all—and after the introduction of radiotherapy, Coley's Toxins eventually fell out of favor.

Fast-forward to the early 1980s, when a researcher—another surgeon, named Steve Rosenberg (Chapter 13)—was heralded for treating cancer with a drug called IL-2, a drug that is natural to the human body and is a critical component of the immune system. Using massive amounts of this substance, Dr. Rosenberg was able to cure a number of patients with a variety of tumor types. Unfortunately, however, the treatment was highly toxic and, like Coley's Toxins, only effective in a limited number of patients. And again, the precise way the drug worked was largely unknown.

For years thereafter, the field of IO languished.

Then came "ipi."

In 2011, a drug called ipilimumab ("ipi" for short) became the first IO agent approved by the U.S. Federal Drug Administration (FDA), thereby setting off the current IO revolution. In the pivotal clinical trial that led to ipi's approval, patients with advanced melanoma—patients with only months, if not weeks, to live—were surviving for years after being treated. In describing some of these patients, oncologists are now even using the word "cured."

In 2014, two more IO drugs were approved: nivolumab and pembrolizumab. One of the patients to receive the latter drug—a patient that otherwise would almost certainly have already died of metastatic melanoma without this treatment—is former President Jimmy Carter. As of this writing, Carter is alive, well, and tumor-free.

This is not hyperbole. This is real.

Unlike previous attempts at IO, scientists know exactly what these two drugs are doing and, in general, that knowledge has been put to work discovering other agents and approaches that enhance the patient's immune system.

This is just the beginning. IO is here. Many hundreds of patients have already had their lives extended using this new therapeutic approach. Very soon, that number will be in the tens of thousands.

Overview

"Behind the scenes they just whispered, 'Ah, I don't believe it.'"
—DIMITRI GABRILOVICH, PH.D., PROFESSOR OF TUMOR IMMUNOLOGY, THE WISTAR INSTITUTE

The IO revolution almost didn't happen. The foundational idea that the immune system could even see a cancer cell, let alone kill it, was to many an anathema. Prominent researchers—highly intelligent men and women—gave the idea a great deal of thought and concluded quite simply that the underlying scientific principles were flawed, and that the approach would never work.

Coley's Toxins had failed. IL-2 was too toxic. Cancer vaccines that made perfect scientific sense on paper were relentlessly ineffective in the clinic. There were some very dark opinions about this technology and for some very good reasons. By the mid-1990s, the anti-IO bias was so strong that researchers who were actually making substantive progress in their work could not convince their peers that their data was real.

"Michel [Sadelain; Chapter 17] pulled me into his wonderfully cramped conference room and showed me the data that was emerging, and I almost fell off my chair. My first reaction was that it was probably not true."

—JOSE BASELGA, M.D., PH.D., PHYSICIAN-IN-CHIEF, MEMORIAL SLOAN KETTERING CANCER CENTER, NEW YORK

But the data kept coming, and among those that were previously skeptical there were some converts. They remained doubtful, but at least they were listening, and waiting.

They were waiting for the clinical trial results.

When the first of the IO drugs finally emerged from animal testing and made it to the clinic—a drug called tremelimumab—some patients (a very few) began to respond. It wasn't great, but it was something. However, the drug had some serious potential side effects, and as the clinical trials progressed the general lack of efficacy became clear. The company developing this innovative IO agent pulled the plug, and that was that.

Yet at almost the exact same time another company's IO drug, a drug of very similar design to the one that failed, was also in clinical trials. Preliminary results for that drug, ipilimumab, were also iffy, but ipi had three distinct advantages.

1. The person who discovered ipi was a highly persuasive, charismatic man who would not take no for an answer.

2. The clinician to first use ipi in a clinical trial had the rare ability to listen closely to his patients—in particular, to one patient who, despite having test results that showed the drug was failing, *said he felt better.*

3. And finally, ipi had a corporate champion: someone within the pharmaceutical company realized the drug was actually working, but he and his team would have to utterly rethink and rewrite the gold standard criteria used throughout the world to verify statistically that a cancer drug was

actually effective. It was a monumental argument to be made, and they made it.

In short, the IO revolution is the doing of some very special people: a determined bunch, if not actually fanatical, because they had to be—because no one else believed in them.

This book is their story.

Based entirely on interviews with the investigators, this book is the story of the IO pioneers. It is a book about failure and resurrection, redemption and success. It is a book about science—about discovery, intuition, and cunning. It is a peek into the lives and thoughts of some of the most gifted medical scientists on the planet.

What this book is *not* is a science textbook. Nor is it, frankly, a rigorously vetted record of exactly who did what, when, where, and how, crammed full of supporting citations (though there is just a bit of that). Rather, this is a life book. This technology is saving, and will save, hosts of lives. Therefore, this book is a celebration of the living—the living, breathing, thinking, charming, arrogant, funny, obstinate, spiteful, joyous, drinking-too-much, not-drinking-nearly-enough stellar human beings who have dedicated their lives to making cancer immunotherapy happen.

Finally, the book is not just about IO. Along the way of the story's telling there will be issues raised—problems in the scientific community like gender, politics, and funding. There will be anecdotes—nuggets found along the road like "Tales from the Dark Night" (advice on how not to give up when it's all going wrong), or a discussion of art history, or the Six-Day War, or Stalin, or Les Paul guitars, or dolphins or chickens or *Star Trek*.

It is a book about people—a surprisingly small group of people who are, in fact, a very tight, fiercely bright, packed-with-passion family.

They just happen to be scientists.

⚜ ⚜ ⚜

Reading Notes

- This is not a single narrative. The technologies of cancer immunotherapy are greatly varied, and the stories of their discovery are often as unique as the technologies themselves. Some discoveries may indeed overlap with others in time, space, and personnel, but many others do not. Chapters set off by themselves are just that: stand-alones that tell a single story. As such, these unique contributions can be read in almost any order.

- There is one chapter that does not read like the rest: the story of Ralph Steinman (Chapter 10). This singular exception to the book's overall design is because, of all the scientists highlighted herein, Dr. Steinman is the only one no longer with us. Although there are many deserving IO innovators that could have been included posthumously, when the idea for the book was first conceived it was decided that the telling of the stories would be left to those who are still here to tell them. That said, Dr. Steinman's contribution to the field is so central to so many of the other technologies described that his exclusion from the book would have been unconscionable.

- There are hand-drawn illustrations facing most of the chapter title pages. Many of the scientists profiled for this book drew these cartoons (that's what scientists call such diagrams: cartoons).

  The images are not meant to be particularly instructive or even dutifully detailed, as they were done on the fly—most were executed in front of me in a matter of minutes—and were not intended to further my personal scientific education. They are simply a quick peek of what was in the scientist's mind at the time when the cartoon was requested.

  Please look upon them as fan autographs, if you will, gifted to me by my favorite stars.

CTLA-4

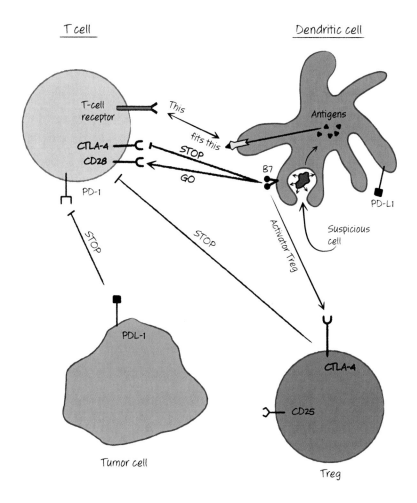

JAMES ALLISON
"Raising the Tail"

CHAPTER ONE

James Allison, Ph.D.

Professor, Chair of Immunology
M.D. Anderson Cancer Center
Houston, Texas

"I proposed treating cancer by ignoring it." —J. ALLISON

James P. Allison was born in 1948, in Alice, Texas.

"Alice—it's a very small town," intones Allison, in his comfortably worn Texas drawl. "It's got tall boots and mesquite, and cactus, and a lot of cows. Maybe it's hard to find on a map if you don't know … It's near Palito Blanco and Freer, if that helps."

A nice place to grow up, but Alice is a strange place to take a shine to science. "I was lucky," Allison is quick to acknowledge, "My dad was a country doctor, so through him I got to see medicine and science, and I also had some pretty good schoolteachers that recognized something [in me]."

These early champions got Allison into special academic programs, and he would spend his summers from the eighth grade on in one science program or another at the University of Texas at Austin. The teachings, and the teachers, made their mark on the scientist-to-be.

"There were two teachers, actually," Allison recalls. "One was Ernestine Glossbrenner. She was my eighth grade algebra teacher and she was very supportive." The other teacher provided both positive and negative reinforcement: "Larry O'Rear. He was my physics and chemistry teacher, except he was complicated because he was a Church of Christ lay minister, and he made absolutely sure that no evolution was taught in the school."

This set up an intractable conflict for Allison. "I'd learned about evolution on my own, and since they didn't teach it in biology class I refused to take high school biology." This decision did not sit well with the school board. "It caused quite a rile," says Allison. "But I told them, teaching biology without evolution is like teaching physics without Newton; I don't see how you can do it. So, they came around." Allison was allowed to take biology by correspondence from UT Austin.

Defender of the Faith

Years later, Allison was called upon again to champion the cause of science education: "So, by this time I'd finished my Ph.D., done a postdoc, and had come back to live in Austin and I get this call." His old eighth grade math teacher, Ernestine Glossbrenner, was now in the Texas legislature and serving on the Education Committee, and she had a problem. "She said there's this crazy guy named Mike Martin who introduced a bill to require teaching of Creation Science in the schools, and you've got to come down and help."

Representative Glossbrenner remembered Allison's run-in with the school board and hoped that he would be willing to stand once again in defense of science. He accepted. Allison would debate Martin in front of a committee of the Texas legislature.

"Martin started in with stuff like, 'If you put a Ford in a field it just rusts, it don't turn into a Cadillac.' That was the level of discourse. So my attitude was, okay, Mr. Martin, you tell me how you can use your creation science to explain how bacteria become resistant to antibiotics. You use your creation science to tell me how a tumor cell escapes the body's immune system. Use your science to explain *anything*. Tell me, give me an example of what you can use your creation science to predict, because science isn't about an incomplete fossil record, it's about predicting things."

As the debate went on Martin tipped his hand to his real concern: There was a secular humanist conspiracy to suppress creationist thought.

Allison bristled, and drove his point home. "I said, no, creation science lost out in the free marketplace of ideas because it's not useful."

He then turned the tables on Martin, pointing out how others in the past have tried to contort science for political or religious reasons. For instance, said Allison, for many years the Soviets emphasized Lamarck over Darwin because Lamarck advocated the inheritance of acquired characteristics, which is more consistent with the socialist/Marxist idea of the perfectibility of man.

"Martin got so flustered by what I said that in his rebuttal time all he could do was keep denying he was a communist. Luckily, I won the debate, and they killed the bill. It was a lot of fun."

Scientist, Know Thyself

From eighth grade on, Allison knew he wanted to be a scientist. "My dad still wanted me to be a doctor, so when I started college I was pre-med," says

Allison. But it didn't last very long. "It quickly dawned on me that the pressures of making decisions, day-to-day decisions that affect people's lives and, you know, you've got to be right. You can't be wrong." A scientist, on the other hand, is expected to be wrong most of the time. That's intrinsic to the journey; most experiments fail. "As a scientist you only have to be right sometimes. I liked that a lot more."

The choice of what sort of scientist to be was not quite so direct, but Allison again let his nature guide him: He likes puzzles; he likes taking things apart. "I was actually trained as a biochemist, not as an immunologist, but I just got interested in immunology." He was fortunate enough to encounter his third mentor. "As an undergrad I took a course taught by a very good, very charismatic professor named Bill Mandy."

T cells had recently been discovered, but Mandy himself didn't believe they were relevant. "He liked B cells; he was an **antibody** guy all the way through." (Antibodies come from B cells: see below.) Nevertheless, the entire topic was compelling to Allison. "The idea that you could have these cells going around in your body, traveling through the lymph nodes, communicating with the other cells and tissues in your body and protecting you from most anything that comes along, even things that might not have existed before, and then somehow do that without killing you? I just thought that was a fascinating biological issue."

The Journey to Ipilimumab

After completing his training, Allison began his research in earnest as a faculty member for the University of Texas M.D. Anderson Cancer Center, initially working out the protein structure of the T-cell antigen receptor

Antibodies are large proteins generated by the B cells of the immune system in response to recognizing a given *antigen*. Antibodies are extremely diverse in the range of what they can recognize. In fact, B cells are able to manufacture antibodies in well over a billion "flavors." Each "flavor," each antibody, is very specific and usually will only recognize and bind to one particular type of antigen—not unlike finding the single archnemesis in a vast and highly varied mob.

An *antigen* is any distinctive tiny aspect—organic or otherwise (it could even be plastic)—that an antibody can recognize. A prime example of an antigen is pollen. Products said to be "hypoallergenic" contain very few antigens, and as such should be ignored by your immune system.

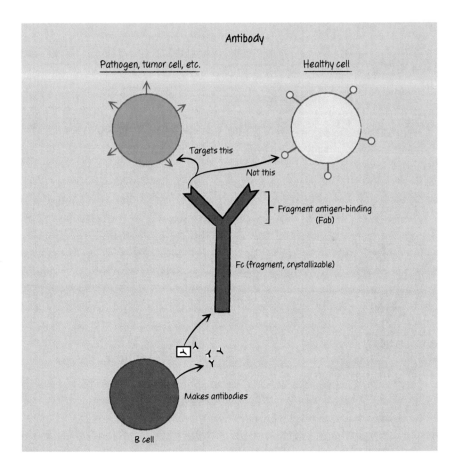

(TCR; see Mak, Chapter 11). "That's the switch, the ignition switch that turns on T cells," he explains. When a T cell encounters an antigen that matches up with its TCR, the TCR activates. "I was interested in general in how you regulate T cells," those cells being the assassins of the immune system. "How do you turn them on? How do you stop them?"

"I was interested in general in how you regulate T cells. How do you turn them on? How do you stop them?"

Ten years and countless experiments later, Allison came across a second activating pathway, a co-stimulatory signal beyond the coupling between TCR and antigen that is essential for an effective immune response. He thought of it this way: If the TCR–antigen interaction is the ignition, then this second signal was the accelerator, revving the engine, driving the T cell onward to fully engage (and kill) its target.

"It was a big mystery as to what that signal, that molecule was, but it was known to be present on very specialized cells called dendritic cells." (See Steinman, Chapter 10.) These are the cells that tip off T cells to the presence of a tumor antigen. The tumor doesn't do this; the dendritic cell does. The co-stimulating moiety Allison was working with is now known as CD28 (cluster of differentiation 28).

"We messed around with that a little bit," says Allison, "but when we cloned CD28 we encountered another molecule that had already been discovered called CTLA-4" (cytotoxic T-lymphocyte–associated protein number 4—i.e., the fourth one discovered). Little was known about this molecule at the time except for a few tantalizing clues: It was not produced in resting T cells, but only in T cells that had been activated, and CTLA-4 appeared to bind the same ligands as CD28: the ligands known as B7-1 and B7-2 found on dendritic cells. A competing laboratory demonstrated that the CTLA-4 receptor bound the ligand more tightly than did CD28. Given this relationship, that group proposed that CTLA-4 was another co-stimulatory molecule.

> Ligands *are the mates of receptors—like two good friends—and like friends, when ligand and receptor get together, they do something. Generally, when a ligand (often in the form of a protein) binds its partner receptor (often located on the surface of a cell) this coupling turns some process in the cell on or off.*

"[The other laboratory] was working in human cells. We were a little behind, but we cloned the mouse gene and made antibodies to the gene product," says Allison. Interestingly, this work was being mirrored by Jeff Bluestone at the University of Chicago (Chapter 21). "Both Jeff and I independently came to the conclusion that CTLA-4 was not another gas pedal, if you will, but was actually a negative regulator in opposition to CD28: a brake."

Further investigation indicated that the other group had misinterpreted a key observation: They had concluded that the antibody they were using was an agonist, based on an observed increase in T-cell activity, "But really what it was is that they were blocking the negative signal." Thus, instead of stimulating new activity, the antibody was restoring existing activity by blocking an inhibitory effect.

Aha!

"The "aha" moment, at least with respect to cancer, came after I started thinking about how tumors just can't give that second [i.e., activating] signal."

Allison reasoned thusly: The immune system has a number of built-in mechanisms to prevent autoimmunity: the attack on healthy cells by the immune system. One such mechanism is cross-priming, the process whereby the cellular debris of a dying cancer cell—DNA and all—induces an inflammatory response that summons cells of the immune system to clean up the debris.

Discrete bits of the cellular debris—the unique tumor antigens—can then be processed by the antigen-presenting cells, like dendritic cells (DCs), and "presented" (i.e., displayed) on their cell surface for recognition and targeting by T cells. T cells that recognize the specific tumor antigen being presented will bind to the dendritic cell where the second signal is then given, stimulating the full-blown immune response.

Note: The activity of DCs is more commonly related to the process of toleration—to avoid autoimmunity—not T-cell activation. Thus, the efforts of many in this book are directed at the more difficult task of ramping up the activation of T cells.

"Once fully activated, the T cell will kill, and continue to kill without further instruction. That's how it works," says Allison. Yet, this proclivity to kill strongly implies that there is an intrinsic mechanism in T cells that can at some point end the carnage, because an unbridled immune response can kill you.

But what was the mechanism?

"Everybody had been concentrating on this process where you get the T-cell receptor signal, and then a co-stimulatory CD28 signal, and then this whole cascade of cell cycle progression and expression of cytokines: all these positive things," says Allison. "But what wasn't realized, what even I didn't realize for a little while, was that this all starts a negative program as well by inducing the CTLA-4 gene, and that's what's going to eventually turn the system off." CTLA-4, it was theorized, serves as a "checkpoint" to limit the immune response.

Evidence supporting this "off switch" checkpoint hypothesis was provided by a rather simple experiment. "We knocked out the CTLA-4 gene in a mouse and found that without it these mice die when they're about three weeks old. They just fill up with T cells because they can't stop an immune response."

Based on these observations, Allison thought, what if activated T cells could actually detect tumors, but the tumor cells themselves were capable of suppressing an otherwise robust immune response? The next logical step for Allison was to try to remove this inhibition. "I figured, let's just disable the brakes by making an antibody that prevents CTLA-4 from binding its ligands, and then we can just keep the immune system running as long as we want."

It worked: An aha! moment built of incremental progress, but an aha! nonetheless.

Nearly Lost in Translation

At the aha! time, however, Allison was not working on the problem of eliminating tumors. "I always wanted to do something about cancer. I've lost a lot of family members to cancer, and I've had prostate cancer myself, but that wasn't why I was doing these experiments. I was doing these experiments to learn how the T cells work, and only after that did I ask the question, 'What have we learned that we can use to treat disease?'"

Allison had learned that T cells could be activated fully through precise mechanisms to kill tumor cells until such time as they are instructed not to. Tumor cells on the other hand, via CTLA-4 signaling, had the ability to instruct T cells to stop the attack. Therefore, clinical translation would be simple: It didn't matter what kind of cancer it was. It didn't matter what the antigen was. All one needed to do was release the brake by inhibiting the CTLA-4 checkpoint.

It was a provocative idea and it was not well received. "Ever since Nixon declared the 'War on Cancer' and the DNA sequencers came along everybody said, 'We're going to sequence, we're going to learn everything about cancer cells, and we're going to beat cancer by learning what causes it,'" says Alison. This translated to the therapeutic revolution of so-called "targeted therapy" and at the time, the targeted approach was considered to be the way to the Promised Land. "And here I was saying you don't need to characterize every cancer cell, you don't need to know what causes cancer. The immune system doesn't know if it's a kidney cancer, lung cancer, prostate cancer; the immune system doesn't know if it's caused by RAS [a mutated protein] or mutant epidermal growth factor receptor or anything. It just knows it shouldn't be there."

"The immune system doesn't know if it's a kidney cancer, lung cancer, or prostate cancer ... It just knows it shouldn't be there."

The second radical notion suggested by Allison's approach was not to treat the tumor directly at all. "I proposed treating cancer by ignoring it," says Allison, proudly. "I said, instead, treat the immune system. That was the idea: just let the immune system rip." In other words, remove the inhibitory factors and allow the immune system to finish its job (with the caveat that the immune system knows the cancer is there, an issue addressed in other chapters).

It was a simple idea, with data to back it up, but there was still a lot of convincing to do. Allison spent the better part of two years making the rounds of pharmaceutical and biotech firms, and the negative feedback was largely the same. First, that this drug was not a small molecule, which would be greatly preferred by drug companies (antibodies are huge and expensive to make), and second, this drug was a form of immunotherapy, an approach that had been disqualified by previous attempts.

Counterarguments were presented. There was a lot of fruitless back and forth, and a near miss: "There was a preexisting patent that Bristol-Myers Squibb had filed," Allison recalls, "But they got the biology backwards: they said it was a positive [i.e., activating] molecule." Allison had the mechanism right and developed a sound intellectual property position from that perspective, but it was still an uphill battle to get anyone to take notice. "Finally, this little company called Medarex expressed interest. They had a mouse that had some immunoglobulin genes replaced with human and so they could make totally human antibodies from the start, so I said okay."

Note: "Humanized" antibodies are produced by genetically engineered mice and can be safely administered to people.

A Phase I trial was performed with humanized CTLA-4 antibodies. Typically, a Phase I study only generates data regarding dosage and toxicity of the treatment, not the treatment's efficacy. "Well, there were three objective responders in that trial, and one of them was a patient on that trial I met during her 10th annual checkup at UCLA after being cured. She's now 14 years out." Allison smiles as big as a Texas sky.

A Rose by Any Other Name, and the Latest Gig

How do they come up with these drug names? No one knows. "When we first started working on it, it was called MDX [Medarex]-010," says Allison. Then, for reasons not given, the FDA named it ipilimumab. "It was kind of a letdown. I mean, I was at Berkeley at the time, so I suggested they at least put an 'H' in front of it—you know, *Hippi*-limumab—but I guess they didn't think that name had enough gravitas for a cancer drug."

Although Allison failed to influence the FDA regarding the drug's new name, he had better luck in naming his blues band: It's called, appropriately, The Checkpoints. "Everybody in the band is an immunotherapist," says Allison. "You know Patrick Hwu (Chapter 15), head of melanoma at M.D. Anderson? He's the keyboard player, and Tom Gajewski, from the University of Chicago (Chapter 24) is the lead guitar player and he really holds the band together." Other band members include lead singer, Rachel Humphrey (Chief Medical Officer, CytomX Therapeutics), drummer Dirk Spitzer (instructor in the Department of Surgery at Washington University School of

Medicine in St. Louis), and on bass, John Timmerman (Associate Professor of Medicine, David Geffen School of Medicine, Los Angeles). Jim Allison plays the harmonica.

"We play every year at ASCO (American Society of Clinical Oncology), and we play at the Society for Immunotherapy of Cancer meeting too. The last three years we played at the House of Blues in Chicago and sold out the room."

Not too shabby for an inquisitive boy from a little town called Alice.

The Checkpoints, the Society for Immunotherapy of Cancer's house band, performing on June 4, 2017. Band members (*left to right*) Ferran Prat, M.D., Ph.D.; Jason Luke, M.D.; Tom Gajewski, M.D., Ph.D.; Rachel Humphrey, M.D.; Jim Allison, Ph.D.; John Timmerman, M.D.; Brad Reinfeld; and Patrick Hwu, M.D. Heard but not seen is drummer Dirk Spitzer, Ph.D. (Reproduced, with permission. © Society for Immunotherapy of Cancer.)

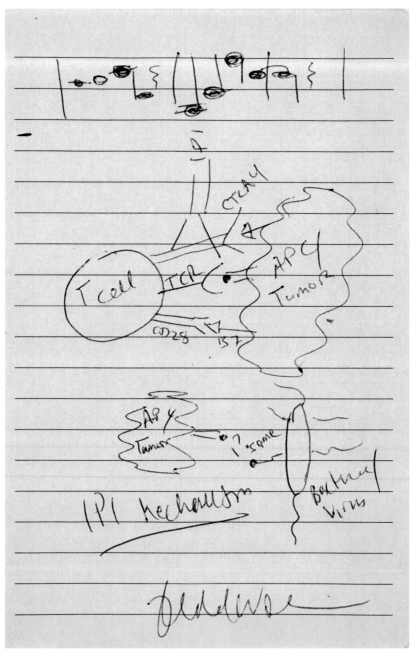

JEDD WOLCHOK
"Ipi"

CHAPTER TWO

Jedd Wolchok, M.D., Ph.D.

Chief, Melanoma and Immunotherapeutics Service
Memorial Sloan Kettering Cancer Center
New York, New York

"Everybody thought we were nuts."   —J. WOLCHOK

And from Dr. Wolchok's mentor:

"In the late 1970s one of my close friends described immunotherapy as treating syphilis before the discovery of antibiotics."   —A. HOUGHTON

Jedd Wolchok, one of the world's top researchers in cancer immunotherapy, was born in 1965 in Staten Island, New York.

"Really," says Wolchok, flashing his ever-ready mischievous grin, "Staten Island. I know. Very few people can say they were born there, and fewer still would admit it."

Note: Staten Island, one of the five boroughs that make up New York City, is home to the now-closed, 2200-acre Fresh Kills landfill, formerly one of the largest dumps in the world.

Fresh Kills jokes aside, Wolchok's done pretty well for himself since leaving the humble island of his birth. He currently serves as Chief of the Melanoma and Immunotherapeutics Service at Memorial Sloan Kettering Cancer Center (MSKCC) and holds the prestigious, MSKCC-based, Lloyd J. Old Chair in Clinical Investigation.

The Mentors

Dr. Wolchok did not cross the waters to Manhattan entirely under his own steam. Like many, if not all individuals of great accomplishment, his abilities were stoked early on, starting with Lesson One: Success means hard work.

"My dad was a Teamster official by day and by night he taught at New York City Community College in Brooklyn: Labor Economics," says Wolchok. "So, I'm no stranger to people who work all the time." Even in retirement, his father, Harold Wolchok, still hosts a radio talk show on car repairs—an outgrowth of his work with the Teamsters.

Wolchok's mother, Elaine, worked her whole life as a New York City elementary school teacher (a phrase that in itself is too exhausting to consider).

And then there was his uncle: "My mom's brother, Irwin Jaeger, was for 45 years probably one of the best AP biology teachers in the country. He tutored me, advised me, inspired me ... What I love about science I love because of what he taught me." As Wolchok tells it, Jaeger had everything it took to be a physician, but he decided to go into teaching, which turned out to be Fate's greater goal.

"He produced hundreds of doctors in his biology teaching career, literally hundreds. I've met them. They are amazingly accomplished people." One of them is David Ginsburg, a highly respected bone marrow transplant specialist at the University of Michigan and a past president of the American Society of Clinical Investigation (ASCI). "I met him recently at my own ASCI induction," says Wolchok, "It was a wonderful reunion."

Mentors, Part Two
The Infancy of Immuno-Oncology

Wolchok's study of immunology could not have come much earlier in his career. While still in high school, he spent a summer working in an immunology laboratory at Cornell. In his first year of college, he met Lloyd Old (1933–2011), one of the founding fathers of the immunotherapy field. Dr. Old introduced him to Alan Houghton who was then, in 1984, the Chair of Immunology at MSKCC.

At the time, aside from scientists at the NIH, investigators in immunotherapy still resembled the archetypal inventor: the lone, fixed-eyed thinker with his late-night tinkering.

"I was 19 years old when I came to work here," says Wolchok. Dr. Houghton had just received tenure and was working on a monoclonal antibody that recognized a target on melanoma cells; he was also the principal investigator of the Phase I clinical trial using the antibody being developed. "So by day he would be in the clinic watching these [hospitalized] people get infused ... and then samples from those patients were brought back to the laboratory. My job that summer was to design the pharmacokinetic assays to measure the amount of antibody in patients' blood over time."

At 19, in his first year in college, working on a clinical trial with actual patients. Right. "If you tried to have a similar conversation today about a Phase I drug in a clinical trial," says Wolchok, "That drug would be made by a biotech or pharmaceutical company, the patients would be treated largely as outpatients, and the drug would be made in some fancy manufacturing plant, and not at an outpost of Sloan Kettering in Rye, New York." And, most importantly, the pharmacokinetic assay would be designed, and its performance interpreted, by a team of highly qualified professionals who would be paid an inordinate amount of money to do this by someone well outside the research institution. In short, it would not be handled as a summer school project by a 19-year-old college freshman.

Yet, it was that experience that motivated Wolchok to devote his life to exploring what was then the nearly black box of immunology. The possibility that the contents of the box might never be revealed or that the revelation would serve no purpose in oncology never entered his mind.

"I have to tell you, that summer, it crystallized the rest of my life because I saw how science and medicine intersected … and since I was surrounded by [immuno-oncology] from that earliest point in my development I knew that it could work. This idea that it couldn't be done, it was never real to me."

"This idea that it couldn't be done, it was never real to me."

As for Dr. Houghton, "He became my most important mentor," says Wolchok, his voice noticeably softer as he speaks of him. "He was chair here for many years, as well as head of the melanoma service [as Wolchok is now] and did all that while he was fighting ALS."

Diagnosed some 20 years ago, Dr. Houghton's condition worsened to the point that in 2007, the leadership position of Dr. Houghton's laboratory was handed to Dr. Wolchok.

The promotion was a sudden, huge step up in responsibility. "It *was* thrilling, but I was extremely content to have my own little operation under his big tent," says Wolchok (he of no discernible ego). "I mean, it's very exciting to lead. It forces you to have some additional vision and develop other skills. But in 2007 is that what I wanted to do? No. Is it what I *had* to do? Absolutely. I mean, there were 20 people working here that weren't going to have a job the next day if I didn't."

And Then Came Ipi

One of the more striking parts of the IO story is just how small a world it once was, and in many ways still is.

"There was a senior scientist working here," recalls Wolchok, "A Ph.D. scientist named Polly Gregor, who had met Jim Allison [Chapter 1] at a meeting and said, 'You know, this guy Jim, he's got an antibody to this molecule CTLA-4 [ipi] and maybe we should start studying it as a way to increase the activity of the DNA vaccines we've been working on.'" And that's how it got started.

Shortly thereafter, Jim Allison himself came to work at MSKCC and, in collaboration with Dr. Wolchok and the corporate sponsor Medarex (later acquired by Bristol-Myers Squibb [BMS]), the first clinical trial with ipi was opened.

At the initial investigator's meeting in 2004, Wolchok remembers being both extremely hopeful and equally wary as he considered the path forward.

"I was sitting there thinking that this medicine is very powerful. It has the potential to modulate the immune system in a very meaningful way, but the side effects that we were beginning to learn about were downright frightening." One of the more troubling ipi-related adverse events was grade 4 colitis, the most serious (to have grade 5 colitis means you died of colitis). "We were instructed to recognize these events early and to intervene in very consistent and specific ways, and I took that really seriously."

But there was another problem unrelated to toxicity and that was the mechanism underlying the efficacy. Ipi is not chemo; ipi is not a tumor-targeting agent; ipi does not act on the tumor at all—ipi acts on the immune system. There are standardized methods in place for measuring the response rates to cancer treatments—any treatment, be it chemotherapy, radiation, or molecularly targeted therapy—called the RECIST criteria. RECIST (response evaluation criteria in solid tumors) measures the rate of tumor shrinkage over a given time period, which—per written protocol in the first ipi trials—was 12 weeks.

But a patient's response to checkpoint immunotherapy is different. It's slower. If it works, RECIST will miss it.

Why Aren't You Dead?
(Aha!)

One of Wolchok's first ipi patients was a gentleman with metastatic melanoma (all initial trials with ipi were in this disease setting). The patient received the drug according to protocol. The patient went home. The patient returned at the prescribed time—12 weeks later—to determine if his tumors had regressed. They had not. In fact, his scans looked worse.

Note: This gloomy radiographic determination was due to something called pseudoprogression, a phenomenon related to the use of checkpoint inhibitors whereby the tumor looks bigger after initial treatments because the tumor is swollen with infiltrating T cells.

Not knowing about pseudoprogression at the time—such a thing had never been seen before—Wolchok informed the patient that, sadly, they'd done all they could do, and sent him home.

After parting ways, however, Wolchok kept thinking about something the patient said just before leaving: Despite his bad test results, *he said he felt better.*

"I didn't know, but I just had this clue that if he felt better, he was going to get better." And even though hopefulness is something of a Wolchok trademark, it was still a surprise when the patient returned, very much alive, for his next scheduled checkup. At 12 weeks, according to RECIST criteria, he was dying, yet 8 weeks later his cancer was nearly gone.

"The irony was that in the official study report for the trial that he was on, he was scored as drug failure," says Wolchok. This, despite the fact that the patient went on to live another 8.5 years, and most of that time with no radiographic evidence of disease. "That really haunted many of us in the field, and we realized that there has to be a better way to judge activity."

While this issue was being considered at BMS, roughly the same conversation was taking place at Pfizer, where a competitor to ipi, a drug called tremelimumab, was being evaluated.

"The [ipi] trial that that gentleman was on was reported as negative. Actually, there were two trials keyed to [standardized] response rates. Both failed." In their trial, Pfizer was seeing the same things with tremelimumab that Wolchok saw with ipi, but the clinical team at Pfizer lacked his prescient confidence in their eventual success. They didn't think tremelimumab would ever work and pulled the plug on further development. (To be fair to Pfizer, most companies, and in fact most clinicians, would have walked away.)

But ipi had an edge: a staunch team of advocates at BMS.

"There was a whole team of internal champions who were absolutely committed to seeing this thing across the finish line," says Wolchok. This was not a matter of blind faith in Dr. Wolchok's enthusiastic mind-set, but rather because they were seeing something they had never seen before: Metastatic melanoma patients with just weeks to live at study entry were living disease-free for years after treatment (not many, but some—roughly 20%). The improvement in disease management was without precedent.

"The question was not whether or not it worked," says Wolchok. "We knew it worked for some people, we just didn't know how to measure the term 'worked.'"

The decision at BMS to move forward was a bold one. "They said, we're going to extend these trials and measure how long people live—which was not a matter of weeks or months—in this case, it took a few extra years to get the data."

And it was expensive.

"Years equate to dollars," says Wolchok, acknowledging the substantial price tag (read: risk) associated with drug development, "But it's that insight, that vision, that opened all the doors, because if CTLA-4 never succeeded … , I do not know what we would have left in development in terms of other immunotherapies."

This need to redefine treatment success in the light of IO outcomes led to new criteria for treatment efficacy being proposed by Wolchok and others, a paradigm shift in the way clinicians look at the results of cancer immunotherapy, a perspective which is now in use in many ongoing IO clinical trials.

Vindication
("Bite Me")

Editor's note: To be clear, the highly educated, deeply conscientious, profoundly dedicated men and women such as those profiled here would never in a thousand years be so petty in their pronouncements as to utter the phrase, "Bite me." Nope. Not this bunch. Not a chance in hell.

Hard work is hard enough. Hard work that no one believes in is agony.

"There were a lot of people who were down on the technology," says Wolchok, but only one person that really got to him. "I have a very dear friend who lives in my building—he's a biotech analyst." And this story takes place around 2006 when the ipi data was not looking so good. "I remember pulling up in front of the building—I'm unpacking the car, we've been away for the weekend—and I'm reaching up on the roof of the car and my friend comes past me and he goes, 'You know, *The Street* hates your drug.'"

> "You know, *The Street* hates your drug."   — BIOTECH ANALYST

It was infuriating. "I didn't know which part of it I was more angry about: that this drug should be seen as *my* drug, or that *The Street* should have an opinion on 'my drug,'" says Wolchok, wincing at the thought. "What I really wanted to say was, 'I hope *The Street* doesn't get melanoma.'"

But then, not too long after this exchange came vindication.

"He came up to me a couple of weeks after things started to look good and he says to me, 'You know, you were right all along.' And he says, 'I want to be your publicist,'" recalls Wolchok, garnishing his words with that impish grin. "He says, 'You need to get out on the speaking circuit and become like, a motivator, because you could tell your story about how you were committed and focused and confident, how you just wouldn't quit.'"

Revenge doesn't get much sweeter.

The pivotal trial results for ipilimumab involving Wolchok et al., with lead author Dr. Stephen Hodi, was the cover story for the August 19, 2010 issue of the *New England Journal of Medicine,* one the finest medical journals in the world.

The Good with the Bad

"Right now it's great fun," says Wolchok, because pretty much everyone is on board with IO (at least for checkpoint inhibitors). "And the most fun part of it all is that for the first time since 2000 when I joined the faculty here I can now have meaningful discussions with people who were just given a devastating diagnosis, not about some pipe dream that might work, but about several new things which actually do work."

As of this writing, there are seven approved IO drugs, with several more expected to be approved in the next year. "The fact that we can sit down with desperately ill patients and talk about real hope, real tangible hope..., that's the greatest fun of all."

And the downside?

"The worst part of my job is when those good outcomes don't happen," says Wolchok. "I had to pick up the phone last Sunday afternoon and call a young woman who was sitting in our emergency room here to tell her that the inside lining of her brain was studded with melanoma metastases."

The drug had failed.

"Before the call, I remember saying to my wife, 'It's time for me to be a grownup.' I've got to make this call. I didn't want some emergency room physician who doesn't know this young woman to share this news with her. I was a 100 miles away. I would otherwise do it in person, but that's the worst part of this: ... We're not helping everybody, and we need to work harder."

The Dark Night

This section, The Dark Night, will appear in many chapters, because part of the aim of this book is to encourage the budding researcher by citing the experience of potential mentors. For some chapters, The Dark Night occurs as part of the narrative. In others, as it is here, it skews more toward general advice for the men and women who are steeped in doubt, those shadowed, advanced-degree denizens facing with dread yet another run of this seemingly doomed, disgustingly complicated, incrementally uninformative experiment.

"I remember those days where you just feel like, if I have to do this plasmid prep one more time and get nothing out of the tube of cesium chloride and ethidium bromide, I'm going to hang up my pipettor," says Wolchok. "But I think that's the moment that shapes people's future, because you either see through that and you say to yourself, I'm going to put up with this crap because I *want* to put up with this crap, and appreciate whatever little victories are to be had on the way." Or, you should quit.

And keep in mind that it's not actually the hands-on work that's so important.

"People used to sequence a gene and get a Ph.D. for that. Now you send the sample off in an envelope," says Wolchok. Such a task is merely contract work now. Granted, learning advanced techniques are all well and good, but it's a way of *approaching* a problem that's the goal of a Ph.D. "It's a way of asking and answering questions and finding your way out of that corner that you arrive at when seemingly there are just two walls and there's nothing … you know, there's no bright light for you to see. And being able to, like, look behind you, bend down, look up, see how high the walls are … *go back to the literature and find some other way*. It's the learning of the thought process that's so important."

The Great Escape

Regarding the musical notes on the upper portion of Dr. Wolchok's cartoon (see the page facing the chapter title page): When the burden of practicing investigational oncology gets too heavy, Wolchok sets it aside and picks up a tuba.

"Actually, I didn't pick the tuba," says Wolchok. "I was a trombone player and my band director in junior high school was this wonderful yet formidably sized man who came up to me and said, 'Son, we need someone to play the tuba.' Given the situation, I felt I was in no position to object, but before I accepted he told me something very interesting: He said, 'If you play the tuba, people will dust off the chair for you.'"

Really?

"Yeah." The gesture is a show of gratitude, of respect, especially in circumstances like that of the Brooklyn Wind Symphony, which is a remarkably good all-volunteer orchestra to which Wolchok belongs. "You rarely get denied a seat in a noncompetitive ensemble," says Wolchok, "because the truth is, there are not that many people out there that play the tuba; it's not a common instrument. I mean, you have to be willing to cart around a lot of heavy metal, and I am not a large man … But, that's the story, and I'm happy that I did it."

Music is his way out, and his way home.

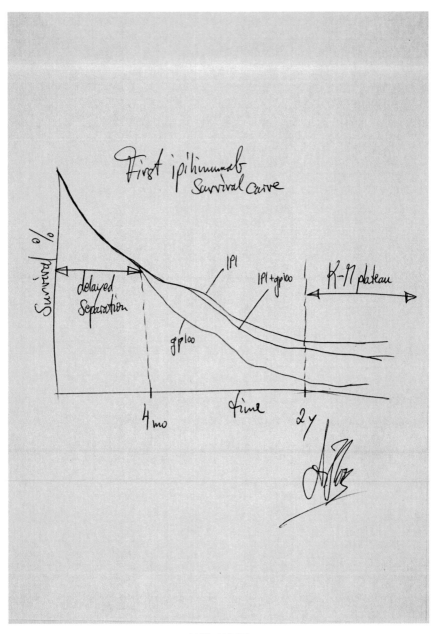

AXEL HOOS
"First Ipi Survival Curve"

CHAPTER THREE

Axel Hoos, M.D., Ph.D.

Head, Therapeutic Area Oncology and Immuno-Oncology
GlaxoSmithKline Pharmaceuticals
Collegeville, Pennsylvania

"This is a nut we can crack."   —A. Hoos

Axel Hoos was born in 1969 and grew up surrounded by the gentle hills and forests of Germany's state of Hessen, in a village called Sondheim, with a population of roughly 200 souls.

"It's a tiny town," says Hoos, in a manner as crisp as his favored blue suits. The town was so small, in fact, that he spent much of his time growing up there in the surrounding forests. There were oaks and birch and there was Hoos, running: long-distance running, which he did competitively and which remains his physical discipline to this day.

"As you train, your ability to perform increases, you have much more capacity to process oxygen, muscles are functioning better, and you produce endorphins when you're moving fast and that feels good," says Hoos, pausing, private man that he is, reaching for the right word to describe what he's probably not put into words before. "You, you feel like you're flying. It detaches you a bit from reality. You can just live the performance."

As a young man, he "flew" through the forests of Sondheim until one day he realized he had to leave. He wanted to do big things. He wanted to make a difference. He couldn't do that in Sondheim. "So I left." Brevity of comment masks the feelings Hoos has for his town and for the quiet of the forest, save for his footfalls. Not one to wear his passions on his sleeve, Hoos is armored by clarity of purpose.

This demeanor is critical to the story of ipi. If James Allison was the wily discoverer and Jedd Wolchok the clinical pioneer, then Axel Hoos was the drug developer and corporate champion.

Making of a Champion

"In high school my favorite subject was philosophy," recalls Hoos, "And German, so ... literature." The draw for both was the same: there was a narrative, and Hoos likes a good story. However, the German school system requires a diversity of interests in their students, so Hoos chose to take biology alongside his core classes. It was a prophetic pick.

"It's interesting—one of the things that excites me in philosophy and literature is stories, human stories, and I quickly found out that science has them too," Hoos says. "You get all kinds of crazy stories: Watson and Crick and *The Double Helix*, for example [the discovery of DNA's structure]; that one really stuck with me." The most important takeaway from that book, according to Hoos, was Watson's conviction. "If you believe in the science that you're doing, and you think you can have an impact, you won't be easily disturbed from that by a nonbeliever—and there are always more of those than not—trying to talk you out of it. And that lesson clearly applies to immuno-oncology. So I got excited about science through these stories that science tells."

Thus, a biology major was born.

But it wasn't enough. Learning for the sake of learning lacked punch; Hoos wanted to make a difference. "I'm always looking to move something, to make an impact." Biology was fine, but the stuff of basic science is generally not geared toward utility. It's not translational work. If you want to make a notable mark in the near term, reasoned Hoos, you take the biology into medicine. So, that's what he did: He chose to be a physician, obtaining his degree from the prestigious Heidelberg University, another prophetic choice.

"On the same campus where Heidelberg University is situated there's a place called the German Cancer Research Center [Deutsches Krebsforschungszentrum (DKFZ), in German]," Hoos explains. "It's one of the largest cancer centers in Europe, with academic science and academic medicine, so I chose to do my Ph.D. there." (Combined M.D./Ph.D. programs do not exist in Germany. Studies are completed independently.)

Next into the mix in the making of Dr. Hoos was the choice of Ph.D. thesis. "I got fascinated with molecular biology," he says—the underpinnings of regular biology. The extra classes with a molecular slant were both a good move and a pain: They meant a great deal more work. "It sparked my interest in a thesis that would be more exciting than a superficial science project." Hoos took his enthusiasm next door to the DKFZ, identifying as his thesis the elucidation of immune regulation by human papillomavirus.

It was a cunning choice. "It was a strong topic at the DKFZ, because the director of the center at that time was Harald zur Hausen." Dr. zur Hausen would go on to receive a Nobel Prize in Physiology or Medicine for his work on papilloma viruses in 2008.

Work on his thesis took three years to complete while Hoos attended medical school at the same time. To sum up: "Basically, a lot of nights and weekends. Little personal time."

The Cutting Edge

All that science still wasn't enough. "I realized I still needed to be a physician," says Hoos, and there was an obvious and profound unmet medical need right there waiting for him. "I had learned about cancer, and I found it a challenging subject because we needed better treatments than giving out poison, because that's what chemotherapy is." If you want to have direct impact on cancer, what do you do? You cut it out.

"The simple solution for me was to become a surgeon and do science on the side." For academics in Germany, it is possible to do both: treat patients and have a lab. At least, that was the plan. "It was a naïve thought. The surgery program is an all-absorbing thing. It eats you."

Not willing to sacrifice the science, Hoos looked for another way, one that presented itself in the form of a research grant from the German government that would cover a two-year stint at an international academic institution. The institution in this case? Memorial Sloan Kettering Cancer Center (MSKCC), New York City.

The cutting edge, indeed.

"I chose Alan Houghton's lab—tumor immunology."

One could almost have said: "Once upon a time . . ."

"I was a fellow in Dr. Houghton's lab [around 2000] at the same time as Jedd Wolchok. That's how we know each other."

The lab was a special place, run by the extraordinary man who had been diagnosed with ALS before Hoos arrived.

"When I first came to the lab, Dr. Houghton had already lost his ability to walk, but he was still able to speak," recalls Hoos. "He had a wheelchair with a joystick. He saw patients, he directed patient care, and he continued to run the lab, running lab meetings, giving advice on how to design experiments. The strength of his mind and his character was so impressive. He inspired people."

During his fellowship, Hoos focused on a technology colloquially referred to as a "gene gun," whereby nanoscale gold particles coated with antigens are loaded into an air gun and "shot" into the patient's skin. It's a method of vaccination.

The work was fascinating. The work was innovative.

The work was not enough. "I was still a few too many steps away from the patient," says Hoos. To move closer, he used his position within the department of surgery at MSKCC to bring scientists and surgeons together. "That's the advantage I had being a surgeon in Alan Houghton's lab. I had access to the entire clinical department." Hoos began offering clinical scientists translational research projects such as measuring biomarkers in a patient's tumor that could be correlated with clinical outcomes, which could possibly be useful for predicting outcomes, targeting new therapies, or potentially help with treatment decisions.

"And that opened the door for many projects because it's translational," says Hoos. You can see the results in the patient. "And that excites everybody." Taking this approach, Hoos and collaborators published 20 papers from the data generated in just two years. Without realizing it, Hoos had assumed his first leadership role, with cross-discipline cooperation as its philosophical core.

"I didn't view myself as a leader," says Hoos. "I just did stuff that led to a result."

The Dark Side

After that stellar two years, the fellowship ended and Hoos wrestled with several questions: Go back to Germany, or stay in the United States? Live in the clinic, or live in the lab? There was a proposal for another surgical residency, this time at Harvard. It was tempting, but he wanted to help patients in other ways. He wanted to do research.

"My choice was ultimately to go into industry, because I thought I could make a greater impact." His reasoning was a bit naïve, but straightforward: "I didn't know what industry was actually all about, I just knew that at the end of the [scientific] story, they are the ones that deliver the medicines."

As luck would have it, Hoos' direct supervisor at MSKCC was Jonathan Lewis, who had recently been offered the Chief Medical Officer position at a Boston-based company called Antigenics (now Agenus). Hoos went with Lewis to Antigenics or, as academics often describe the pharmaceutical industry, "The Dark Side." "'Dark Side,' right? I sometimes jokingly use the same phrase, but I've never seen it as that." What Hoos saw was merely a new path to deliver greater impact.

Of course, this view is not universally held; in any situation where huge sums of money are involved there will be tensions. There will be egos. And there will be attitudes. For the most part though, as one might expect, Hoos just cuts to the chase. Everyone is doing their level best to get something done in the best way they know how. Seeing the world as otherwise is counterproductive, and it's difficult to imagine that Axel Hoos, at any point in his life, did anything counterproductive. "In today's environment everything is collaborative. You need both sides coming together to be successful."

The Road to Ipi

Hoos started at Antigenics with no previous background in industry, and no experience in drug development, "So I had a steep learning curve." Antigenics had a product in development called Oncophage: a personalized vaccine approach using the patient's own surgically resected tumor tissue as a basis for the treatment.

"The approach made good scientific sense," says Hoos, "And it was something that brought several worlds together: the worlds of surgery, molecular biology, immunology, and industry all packaged in one thing, and I liked that, and I learned a lot."

And it didn't work.

"It didn't play out because at the time nobody had heard of checkpoint modulation," which turned out to be a *huge* problem: Any inflammatory activity of T cells caused by vaccines could be attenuated by an immune checkpoint, thereby resulting in insufficient clinical activity. And there was another problem that wasn't entirely clear at the time: The clinical activity patterns of IO therapeutics (e.g., vaccines, checkpoints) are different than that of chemotherapy or targeted agents.

"I was at Antigenics for about five years," says Hoos, "and midway through it became very clear that the development paradigm we were using to develop a cancer vaccine by handling it like a chemotherapy does not make sense."

To address the issue, proactive as ever, in 2002, Hoos started the Cancer Immunotherapy Consortium, a nonprofit dedicated to the systematic improvement of drug development problems associated with cancer immunotherapies. Its first task was the review of clinical endpoints for immunotherapies as observed by the clinicians using them.

Originating from this years-long, cross-discipline, multicenter effort were a number of critical publications. One in particular, written by Hoos in 2007, laid the foundation for a paradigm shift in clinical trial designs for many immunotherapeutics. Two years later, Hoos—along with colleagues Jedd Wolchok, Rachel Humphrey, Steve Hodi, and several others—published "Guidelines for the evaluation of the immune therapy activity in solid tumors: Immune-related response criteria" (*Clin Cancer Res* **15**: 7412 [2009]), a landmark position paper on immune-related response criteria (irRC) that presented a new way of judging treatment response to cancer drugs.

This new understanding was the key. The fledgling IO industry finally had some guidance, the irRC, without which ipilimumab might never have been approved, and the immunotherapy approach might have once again slipped into the shadows.

Ugh

In the meantime, Oncophage failed and Hoos hit the pharmaceutical road.

"I had five years of experience in biotech," says Hoos, ever determined. "I wanted to develop the drug. I wanted to get the drug over the finish line." It didn't happen, but Hoos was beginning to understand why. First, something was turning off the activity of T cells (checkpoint modulators had yet to be named). Second, the response impact of drugs affecting the immune system cannot be easily measured using traditional means (thus, the Consortium above). And third, small biotech did not have the required clout to do these trials. "I learned a good amount there, but I missed something. I needed the Big Pharma experience."

Hoos moved to Bristol-Myers Squibb (BMS).

BMS and Ipi

The early clinical trials with ipi had been reported by Jedd Wolchok and others, and despite low response rates (5%–10%), Hoos had a good feeling that the drug's mechanism would pan out. "Most people were thinking this was never going to work," says Hoos. Besides, targeted therapies were the real magic bullet—everybody knew that—why bother with the impossibly complicated immune system?

This suspension of belief (data be damned) was endemic to the field. A best-selling case in point, as pointed out by Hoos, *The Emperor of all Maladies*, published the year before ipi was approved, contains not one word about cancer immunotherapy. "The information was out there saying that immunotherapy was doing something, but it was ignored. Mainstream oncologists didn't believe in it."

No matter. They believe in it now. And here's why: BMS had partnered with Medarex, and what had been put in place at the time was a drug development program designed by a clinical team skilled in the study of chemotherapy, replete with the standards of the RECIST criteria. It was inappropriate and would have resulted in failure. "[That team] put a chemotherapy plan in place," says Hoos, but to be fair, he adds that all oncologists would have put a chemotherapy plan in place at that time. No one knew better.

But Hoos and crew thought they could do better, and with no small amount of arm-twisting of the internal stakeholders (it took two years), aspects of the new clinical development paradigm were implemented for ipi. "We probably wouldn't have been successful on our own if Pfizer didn't have tremelimumab, another CTLA-4 antibody in the clinic at the same time. So...two CTLA-4 antibodies, both being tested in metastatic melanoma in Phase III programs at the same time." It was a head-to-head race. One failed. One didn't. Why?

It was a matter of timing. "When Pfizer failed, BMS had to rethink what we were doing for the obvious reason: we could also fail," says Hoos. "And if Pfizer failed with a standard approach and BMS had that standard approach ..." Suddenly (after two years), "BMS reconsidered and started using the new paradigm."

Prior to this clinical shift in the interpretation of treatment response rates, the measure of tumor shrinkage (according to RECIST) was taken as a predictor of what might be expected in terms of overall survival, a rate that is always less than the response rate, often much less. This discouraging phenomenon often results from the development of treatment resistance: patients initially respond to treatment and feel better, only to eventually relapse and die.

"Here, it was the complete opposite," says Hoos. "Response outcomes were small and survival outcome was big." Some patients who did not respond showed a tendency to still do very well with survival. Some survived for years beyond their initial short-term, terminal prognosis. Some have been in remission so long as to be considered cured. "So we figured that we did not yet have validated tools to capture all benefit a patient can experience from immunotherapy. We therefore changed the primary end points of our Phase III trials to survival, which ultimately would capture the therapeutic benefit."

The Dark Night

"Changing the paradigm of how things are usually done is hard and it was no exception for ipi. In six years there were many difficult moments that took much resolve to manage. In the end, my plan was I want to have impact. And this is a nut that we could crack. So I stayed with it."

Okay, then.

Vindication

A single moment can settle a lifetime of arguments. It might be an apology. It might be a hug. It might be statistics.

"The very first time I knew I had something big was when my statistician at BMS, Tai Tsang-Chen, came to me and showed me the Phase III survival curve for ipilimumab," recalls Hoos. "Mind you, this is the first person to ever see the data, and he pulls it off the printer and says, 'Oh, by the way, here's the survival curve from that trial.'" And he sets it down.

"I looked at the survival curve" he pauses, with half a lifetime of work filling the silence, "I looked at the curve and I knew I had something." The curves showed that, initially, patients treated with ipi and patients in the control arm had roughly equivalent rates of survival for the first few months, but after more follow-up time—far longer than you would expect with chemo—the two curves, one immunotherapy, one not, began to separate. (See the page facing the chapter title page.)

That was merely the good news. The great, the amazing, the jaw-dropping news was the so-called tail. For chemo, a survival curve starts with a line at the upper left corner of the graph (when all patients in the study are still alive), and then the line progresses downward to the right as patients on the study die off. Because very few cancer treatments are curative, that line, if you wait long enough, eventually winds down to a surviving population of near zero.

That's not what Hoos was looking at. He was looking at a tail, a line, that sloped ominously downward over two years as time progressed, but then plateaued, flattened out. *Finito.* The tail of the curve showed that roughly 20% of patients were showing long-term survival in excess of two years.

> "*Immunotherapy was no longer a hypothesis.*"

Hoos let it sink in, and then he knew: "Immunotherapy was no longer a hypothesis. This was an answer, a definitive answer. It's working."

On March 25, 2011, the U.S. Food and Drug Administration approved ipilimumab for the treatment of unresectable or metastatic melanoma.

The long-distance runner had finished his finest race.

✤ ✤ ✤

As anyone with any knowledge of how drug development in a large pharmaceutical corporation works can tell you, no one does anything alone. There are always teams. That said, were this book to tell the full corporate story of ipi, where all the relevant characters are called out for their particular vital contributions, that would be a tome unto itself. The choice of having the story told through the eyes and efforts of Axel Hoos is justified by the opinion, widely held among his colleagues, that Hoos "has been at the forefront of interactions between regulators, industry scientists, and academia in the wider world of IO." With some of the clinical infrastructure in place when Hoos arrived at BMS—ipi was partnered between Medarex and BMS and some trials were underway—Hoos' pivotal role was as drug developer and champion, adjusting, expanding, and pushing the moving parts of a complex and changing program forward. Nevertheless, credit where credit is most deservedly due. The following is a list of critical players in the ipi story, with their current affiliations.

BMS: The Ipi Team

Rachel Humphrey (CytomX)
Aparna Anderson (Statistics Collaborative)
Tai Tsang-Chen (BMS)
Axel Hoos (GSK)
Ramy Ibrahim (Parker Institute)
Kevin Chin (EMD Serono)
Greta Gribkoff (BMS)
Heather Knight-Trent (Chimerix)
Renzo Canetta (retired)

BMS: The Leaders Responsible for Acquisition of Medarex

Elliott Sigal (New Enterprise Associates)
Brian Daniels (5AM Ventures)
Jeremy Levin (Ovid Therapeutics)

Medarex: The Team That Initially Created Ipilimumab

Geoff Nichol (BioMarin)
Nils Lonberg (BMS)
Alan Korman (BMS)
Michael Yellin (Celldex)
Israel Lowy (Regeneron)

Note: *Lonberg and Korman created the pharmaceutical version of the human antibody to CTLA-4. James Allison's anti-CTLA-4 drug was from mice.*

Clinicians

Jedd Wolchok (MSKCC)
Stephen Hodi (Dana-Farber)

Jeff Weber (NYU Langone)
Steven O'Day (John Wayne Cancer Institute)

Note: *The investigators listed above, who were treating their patients on the very first clinical trials with ipilimumab, provided the critical bedside observations that contributed to the formulation of the new immune-related response criteria.*

To Conclude

"Individual commitment to a group effort—that is what makes a team work, a company work, a society work, a civilization work."

VINCE LOMBARDI

SECTION II

PD-1

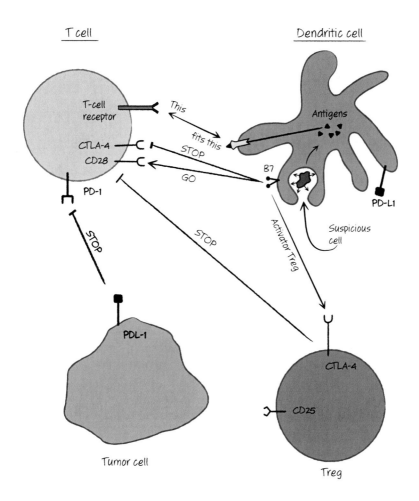

TASUKU HONJO
"Two-Phased: Anti-PD-1"

CHAPTER FOUR

Tasuku Honjo, M.D., Ph.D.

Professor, Department of Immunology
and Genomic Medicine
Kyoto University
Kyoto, Japan

*"Basic research? It's great fun. Great fun! I never imagined [it could be]
so good."* —T. Honjo

Tasuku Honjo was born in 1942 in Kyoto, Japan.

"Kyoto is a very nice place," says Honjo, speaking quietly, slowly, and
finally confidently as he underlines each word with an invisible link to the
supporting data. "It is also a very old place. It was the imperial capital of
Japan for over 1000 years, and so it has lots of historical places. And a river
runs through the middle of the city. It makes the atmosphere … very pleasant."

Born at the height of World War II, the pleasantness of Kyoto was a later
realization. "I spent early life in many different cities. My father was a surgeon
and so we lived in many parts of the country."

That Honjo became a physician like his father seemed an obvious
choice. In fact, there were any number of doctors in his extended family to
serve as role models. Yet, this background was more kindling than spark: "I
had another big stimulation," says Honjo, "I read the biography of Dr. Hideyo
Noguchi, who came to the United States before the war, to the Rockefeller
Institute, and he did very hard work in [the] microbiology field." Honjo pauses
to corroborate his following assertion with the data set in his head. "He made
a very important contribution to medical science, so that stimulated my boy's
dream to become a medical scientist." (The source of that inspiration, Dr.
Noguchi, had overcome serious physical disabilities suffered in a childhood
fire to become a premier researcher. He is most known in his field for unravel-
ing the pathogenic mechanisms of the bacterium, *Treponema pallidum*, the
causative agent of syphilis.)

Inculcated in this way, both by his family's general calling and a Japanese scientist's fame, Honjo returned from hither and yon to begin his medical studies at Kyoto University.

"I was very fortunate," says Honjo, "When I went to medical school I met the biochemist, Osamu Hayaishi, who discovered the enzyme called oxygenase." The discovery was a substantial contribution to the field for which Hayaishi was awarded the prestigious Wolf Prize in Medicine in 1986. "He was a world well-known chemist, very international, and very [proficient] in science. In that way he was very stimulating for us as students."

Honjo found a place in Hayaishi's sprawling lab, despite not wanting to pursue his mentor's principal investigations. "I didn't study his field, but he allowed young people to carry out whatever they [wanted]. He didn't enforce any specific project. So that was a very special environment."

In that environment, Honjo learned what it takes to be a good scientist. Item one: Curiosity. "If you don't have any curiosity you just learn," says Honjo, "and that is not science. You must want to *know something*, to want to dig it fresh out from the ground." That's how you discover. The rest is just preparing for exams.

The second axiom writ large in the "Tao of Honjo" has to do with degree of difficulty, because curiosity seeks out candy bars more often than kale. "You must be attracted to something very difficult, and this represents the Challenge." And finally, to face the Challenge you need Courage. "These are the three things fundamental to being a scientist."

Those are the three qualities you need going in. But what about The Dark Night, when everything is going wrong and Courage seems a barren comfort?

"That's the next round," says Honjo. "You start with Curiosity, a Challenge, and Courage, and then you have to continue this journey with Patience. You *must* have patience. And then you concentrate, you devote ... Then gradually you begin to accumulate your confidence," Honjo concludes. "It is not the other way around."

> "You start with Curiosity, a Challenge, and Courage, and then you have to continue this journey with Patience. You **must** have patience."

For example, lately, Dr. Honjo has been investigating antibody diversity in the immune system, and how the activity of an enzyme called activation-induced cytidine deaminase (AICDA) contributes to that astounding variability. At baseline, a healthy person's genetic code accounts for at least 15 million different antibody designs, each one tailored for a different pathogenic target. The mechanism of AICDA further increases that diversity to around a trillion. But how exactly does that happen? (Curiosity.)

"I have a quite different opinion from many people in the field about the mechanism of AICDA," explains Honjo. (A Challenge!) The debate has been going on for more than a decade: What is the precise mechanism for this enzyme? Honjo says he knows. Others say otherwise. (Courage!) "This disagreement is okay, because we are gradually accumulating evidence, and eventually we can prove it completely." (Patience.)

But as Honjo hones his evidence in hopes of one day driving home his investigative point, he always keeps in mind The Question, the embodiment of doubt intrinsic to the nature of scientific pursuit: What if I'm wrong?

"Some people have too much confidence in one particular idea," warns Honjo. "I see people waste a lot of time, so you have to be objective. Always you have to ask yourself whether this is the right way or not."

For younger researchers there can also be rookie mistakes. A lack of control in an experiment is the most common. (For a quick example of a controlled experiment, consider how to determine the therapeutic effect of a new drug: Take two genetically identical mice, give the drug to one, and give nothing to the other. Observe. The mouse that didn't get the drug is the control.)

Even a failed experiment is valuable with the right control. "There are two types of bad experiments," says Honjo. "One type is the design of experiment is lousy, so you cannot extract anything out of the experiment." For example, with no control there's no way to know if the results are valid. "The other experiment type you just *think* is bad because it's against your expectation. But those are actually not bad experiments. Even if it's a failure, if you have a good control, you can extract something."

So, hang in there. Be patient. And don't think your (well-controlled) experiment failed because your results are at odds with those of the published literature. "Young scientists tend to believe published papers, so if they don't get the result that matches with [the literature] they think they made a mistake. But for me it's something great," Honjo smiles, "You find something interesting!"

PD-1: Something Very Interesting

Observe, and keep an open mind. The discovery of PD-1 nicely illustrates the potential graces of this pairing, because Honjo found a treasure he was not looking for.

"It was a fortuitous incident," admits Honjo. Fortuitous, in that his lab was not looking for another negative regulator of the immune system, like CTLA-4, which PD-1—Honjo's discovery—turned out to be. Instead, they were looking for the molecular signal that tells T cells to drop dead, a process called thymic selection.

Thymic selection occurs in the thymus, a small gland located just above the heart, where T cells undergo a sort of security clearance. To wit: The immune system perpetually churns out randomly targeted T cells, which are then herded to the thymus where they are screened before release into the body at large. If the thymic screening determines that one of these randomly generated T cells is targeting self-antigens — the proteins that identify you as you — that rogue T cell is eliminated through a process called programmed cell death, or apoptosis. Bluntly, the cell is instructed to commit suicide. This fail-safe process protects us from autoimmune diseases such as multiple sclerosis, lupus, or type I diabetes, where the immune system, via T cells, attacks the self.

The question was: What protein signal triggers programmed cell death in self-targeting (potentially, lethally so) T cells?

"One of my students was very excited to work on this," Honjo recalls. To facilitate the search, the student established a cDNA library: a vast collection of discrete DNA snippets that represents what proteins the cell actually makes (much of the DNA in a cell is either inactive or serves other functions beyond the manufacture of proteins). Using a method called subtractive hybridization, Honjo and his team revealed the coding DNA sequence for what is now called PD-1 (programmed cell death-1).

A DNA sequence, however, can only tell you so much about the resulting protein's function. One thing it can tell you is where it belongs in a cell, which in this case is on the cell's outer membrane, with the PD-1 receptor protein protruding like an antenna waiting for a signal. "But what we didn't know is whether it's a positive signal, negative signal, or a killing signal, or anything," says Honjo.

They did figure out fairly quickly that the protein was expressed exclusively by cells of the immune system, but figuring out what PD-1 was doing there, what sort of signal it was looking for, took a bit longer. "We [were] very curious to understand [what the mechanism could be], but it took many years because we used the gene **knockout** technique," says Honjo, speaking matter-of-factly of many years as only an older, successful person can.

It took as long as it did because knockouts take time, and further, you never really know what to look for in the resulting mutant animal. Maybe a noticeable, measurable change occurs, maybe it doesn't.

Knockouts: You want to find out the function of a given protein. You identify the genetic instructions for that protein in the cell. You disable that genetic sequence in the germline (eggs and/or sperm) of the organism (in this case, a mouse). You breed the mice and hope that the gene you knocked out doesn't kill the offspring in utero (although that too can be informative), and then you watch as the mouse matures to see how its overall health is affected by this genetic hobbling—this knockout.

Knocking out the gene for PD-1, however, presented a quandary: Unlike the dramatic phenotypes seen with CTLA-4 knockouts performed by James Allison (Chapter 1) PD-1 knockout mice seemed more or less just fine, at least initially.

"CTLA-4 knockout has a very strong phenotype, and the mice are killed," says Honjo. It happens within weeks, whereas PD-1-deficient mice only developed rather mild autoimmune diseases after a period of months (more than three months is a long time for a mouse; it is also a long time for a struggling Ph.D. candidate). "This student was literally crying, and very much disappointed," because for the longest time nothing was happening. "But one day he comes with a big smile. He had a big moment: The mice are developing diseases!"

Patience and careful observation, "That's biology," says Honjo, happy as can be.

You Can't Spell Ipi without IP

Note: At the time of this writing, the intellectual property (IP) rights surrounding the discovery of PD-1 and its ligand, PD-L1, are in litigation. The PD-1 story, as told hereafter, is in no way intended to clarify who discovered what, when, how, or with whom, but retells the tale as was told to me—in fact, this whole book is like that. I leave it to the lawyers to plow through the literature, notes, and testimony to decide who will eventually get their due.

The phenotype of the PD-1 knockout mice demonstrated that the protein was an immune system down-regulator: a braking system, if you will. "And if you destroy the brake, you can enormously augment immune function," Honjo says, which suggests a number of therapeutic possibilities. For one, you augment the activity of PD-1 (slam on the brakes) to stop an immune attack, thereby helping patients with autoimmune disorders. Conversely, you could block PD-1 and ramp up the immune response to serious infection or, as it was hoped, tumors.

It was decided to investigate the use of blocking PD-1 in tumors, and the details of this successful application were published in 2002 (Iwai et al., *Proc Natl Acad Sci* 99: 12293).

"The next difficulty was to find the [pharmaceutical company] partner for the clinical application," says Honjo, and for that you first need to get a patent for the technology.

"At that time Kyoto University [did not have the capacity to apply for patents]." They lacked the staffing to write it, and they lacked the money to file it. Instead, they insisted that Honjo seek an industry partner. "So I [talked] to Ono Pharmaceutical Co. with whom I had [collaborated] on a different subject at the time." Ono agreed to file the patent on Honjo's behalf.

Honjo then asked Ono for help in developing an anti-PD-1 drug. At this point, there was no drug that you could use in people—only a mouse-based antibody, and a validated target. Ono replied that their experience in oncology was insufficient and they required collaborators who were better equipped, which opened up the discussion to include multiple Big Pharma players. This went nowhere; no industrial entity was willing to invest their resources in the project. "They spent almost a year and then they say, 'It's hopeless.'"

> "They spent almost a year and then they say, 'It's hopeless.'"

It was then up to the scientist–academician Dr. Honjo to investigate the high-stakes financing of drug development by himself. "I went to ... one venture capital [VC] group in Seattle, and I presented our findings and they [were] very much interested. They agreed to collaborate." However, the VC wanted Ono out of the picture, which Ono was not so happy about, and which prompted the company to more productive and timely efforts. "After three or four months they [Ono] came back and said they would do it. They did not explain why they changed their mind, but later I heard that [it was] because Medarex—now bought by Bristol-Myers—had discovered our patent [applied for in 2002]."

Medarex had developed the CTLA-4 drug, ipilimumab, because they had the technology to turn a mouse-based antibody used for research into a humanized antibody for human drug development. "So for Medarex, it's [a] very obvious target to work on," says Honjo. "They approached Ono and they started a collaboration, and then it went very quickly soon after."

The anti-PD-1 drug, nivolumab, was approved in Japan for use in patients with unresectable melanoma in July of 2014. Approval in the United States occurred just before Christmas that same year.

❧ ❧ ❧

Simple, yes?
No, not really.

Concurrent with the search for a pharmaceutical industry partner to help develop an anti-PD-1 drug was the search to identify the molecular partner for PD-1: the signal that turns PD-1 on, that activates the brake.

This work was done by Honjo in partnership with several others, including Gordon Freeman, of the Dana-Farber/Harvard Cancer Center (Chapter 5).

And, thus, wrapping up the Tao of Honjo you have: Curiosity, Challenge, Courage, Patience, and Collaboration with your friends.

GORDON FREEMAN
"Block the Off"

CHAPTER FIVE

Gordon Freeman, Ph.D.

Professor of Medicine
Harvard Medical School
Boston, Massachusetts

"It works where nothing worked before." —G. Freeman

Gordon Freeman was born in 1951 in Hackensack, New Jersey.

"But I grew up in Fort Worth," says Freeman, "I remember moving there in the middle of a hot, hot summer." Which is to say, he moved to Texas.

Like James Allison (Chapter 1), Freeman is an immunologist from the Lone Star State. "Well, Jim's from south Texas, Fort Worth is north central," says Freeman, whose soft, soothing voice has not a trace of Texas twang. "What we really have in common is [we] both got our introduction to big league science as high school students by going to a National Science Foundation–sponsored summer program at the University of Texas [UT]."

There was money for science at the time, and lots of it, because this was the mid-1960s—just a few years after the launch of the Russian satellite, Sputnik—and the fear of being scientifically surpassed by a bunch of Godless communists was very, very real. The funding was so comprehensive that it penetrated all the way down to the high school level. "A faculty member at UT Austin started a summer program for bright high school students," says Freeman, having been one of those. "He got scientists there to let us work in the labs during the summer, so Jim and I both got our introduction to high-level scientific research during those summers."

Not that Freeman hadn't previously been exposed to the scientific method: "I went to this church camp one summer," remembers Freeman, "Out of all things we could have done, they had us dissect a chicken, which was the funniest thing for a summer camp to do. There were boys and girls there, and there was one junior high school girl and myself who actually enjoyed taking it apart and seeing how everything was put together."

They didn't eat the chicken afterward, however, because that would just be wrong. "Actually ... I'm afraid we did. Well, let's put it this way, they were served that night," though Freeman recalls that he and his lab partner opted for iced tea and vegetarian options.

It was also in high school that Freeman encountered his first mentor, Flavin Arseneau. "We called him Mister A," says Freeman. "He was a World War II fighter pilot who'd become a science teacher. After the federal government started funding science research he [put together] a high school lab where you could do scientific research." To join the lab you had to pass a test, a really nasty test that consisted of cleaning up after the previous year's class, which meant handling hundreds of unwashed, remarkably crud-laden, lab containers. "We used glass [reusable] Petri dishes in those days, so if you were determined, you spent your first day cleaning up old, moldy Petri dishes, but then you could do whatever you wanted. We got a key to the high school lab and we could come in on weekends or evenings or anytime, and we learned how to do hands-on lab work." The experience was invaluable.

Note: Lieutenant Commander Flavin James Arseneau (1925–2008) flew carrier-based Hellcats during the war. Surviving that, he went on to found the Fort Worth Science Fair and was a pioneer researcher regarding the effects of DDT on marine and wetland wildlife. From his obituary: "Many of his students entered the fields of research and medicine, making our world a better place."

Mentored by Mister A, and tempered by the UT National Science Foundation high school program, Freeman was good to go for a run at higher education. "The work I did [at UT] helped me be a finalist in what was then called the Westinghouse Science Talent Search." With such a solid foundation in laboratory discipline, and the beginnings of a science CV, it's no surprise that Gordon made it into Harvard. "And I've been at Harvard ever since."

The Path with Less Resistance

After undergrad studies at Harvard came the graduate program, and then finally a postdoc, and that's when Freeman chose immunology as his focus, which would not have been an obvious choice to some. "Yeah, there were complicated things like Arthus reactions and a lingo that was like psychology," says Freeman, but the puzzles of immunology were tantalizing, and the new tools to find and fit the pieces together were enabling his choice. "Science goes through periods of what's doable. Like when I [first] was a graduate student, what was doable was understanding the life cycles and strategies of viruses." For instance, if you wanted to study the SV40 retrovirus, the tools to study the three or four genes that constitute the virion were readily available.

As Freeman progressed in his studies, the advent of the field of molecular biology began to fill tool kits with ever more exotic instruments, enabling things like gene sequencing and the subsequent cloning of genes, which could be used to shine light into the blackest of boxes. "So immunology seemed the next doable area."

Cloning, as the term is typically used in science, only rarely refers to the creation of an evil army in a galaxy far, far, away. More commonly, to clone a gene is to chop it out of an organism's DNA (genes are the genetic instructions to make proteins) and then splice that gene into a much simpler biological system. For instance, the bacteria Escherichia coli can be used to manufacture many copies of a protein according to the instructions provided by the gene you spliced in. With a bucket of identical proteins in hand, you can then start investigating the protein's function.

B7: All in the Family

For his postdoc training, Freeman joined the lab of Lee Nadler, who at the time was working to identify members of the so-called B7 family, a group of structurally similar molecules that were somehow related to the stimulation of T cells. Soon after Freeman joined, the lab discovered the B7-1 and B7-2 molecules. Both molecules turned out to be important because they are the major ligands (i.e., binding partners) for the CD28 receptor located on T cells that, when triggered via binding, activate the T cells. They also turn out to be the ligands for the *immunosuppressive* activity of the CTLA-4 receptor.

Confusing? Well, yes, until you understand that the actions of the immune system are by necessity a juggling of priorities. The first obligation of the immune system is not to kill bacteria; the *first* job of the immune system is not to kill *you*. Therefore, no immune response is intended to be open-ended. Ideally, you want the response to last just long enough to clear the source of inflammation—say, an infection—and then you want it to stop. Human biology uses a rather nifty method of resource management to maintain this balance: The B7 molecules are responsible for both.

It works like this: At the onset of an immune response, the B7-2 ligand (located on the surface of an **antigen-presenting** cell) binds to the CD28 receptor on T cells. This acts like a bugle call, expanding the ranks of T cells and sending them into battle. As the battle rages, the CTLA-4 receptor begins to appear on the field (on the surface of T cells) to discuss the terms of a cease-fire. When enough B7-2 binds to the CTLA-4 receptor on T cells, the call to retreat is sounded.

Antigen-presenting cells, or APCs, are immune cells that exhibit on their surface what are essentially microscopic "wanted posters": molecular thumbnails intended to alert passing T cells to the distinguishing characteristics of a given antigenic "villain." When the T cell assigned to chase that specific villain stops to read the poster, that T cell is then activated via the T-cell-bound CD28 receptor and the APC-bound B7 ligand signaling.

An aside on signaling: If it seems that many of the signals discussed throughout this book are redundant, it's because they are. The reason for this redundancy is no different than why your computer asks you to verify the Delete command you just gave to get rid of that unwanted file. When it comes to doing something that can't be undone, your immune system, like your computer, sets up thresholds of activation so that definitive, consequential deeds are not implemented rashly.

The other thing to understand about the immune system (or just about any biological system) is that they use families of molecules that have very similar activities, and that the members of these molecular families are structurally, and therefore genetically, similar. As with people, if you know what one family member looks like, it's often possible to spot another.

Freeman used this familial relationship to perform a homology search of expressed sequence tags (ESTs): essentially, a fishing expedition in a genetics database for other members of the CD28 family using the known structural and genetic information of CD28 as bait. What he caught was another stimulatory checkpoint receptor called ICOS. Given the relationship to CD28, ICOS was correctly assumed to be in the family business of a receptor that, when bound to a ligand, stimulates T cells.

What was that ligand? Again, Freeman went fishing, this time using the lure of B7-1. This time, however, the genetic angling produced a mystery catch: The ligand that he found did not bind to ICOS. As good fortune had it, Freeman was already collaborating with Dr. Clive Wood of the Genetics Institute of Cambridge, where Tasuku Honjo, discoverer of PD-1 (Chapter 4), had also paid multiple visits.

"That the discovery has led to a revolution in cancer care — that's an incredible thrill."

Using Wood's expertise in creating protein fusion constructs, the connection was made between PD-1 (the receptor) and the mystery ligand, now known as PD-L1.

Note: At this point in the telling of the story, the ongoing litigation with PD-1/PD-L1-associated IP came up, but was very politely set aside. "I think we're all getting a good deal of credit," says Freeman, "And that the discovery has led to a revolution in cancer care—that's an incredible thrill."

PD-1 and Cancer

One of the nicer things about being an immunologist is that your discoveries are quite often translatable, meaning that whatever it is you've got in your test tube could one day wind up in a medicine cabinet.

"When we first discovered PD-L1, we had multiple possible applications," says Freeman. After all, we're talking about a switch here; turning it off means one thing, leaving it on means another. "Say we use it to deliver an off signal; that could alleviate diseases of autoimmunity," like lupus, multiple sclerosis, etc. That's one possibility. Another would be to express PD-L1 on donor cells to tolerize them for transplant. (Tolerize means to force the immune system to leave the transplanted tissue alone, to tolerate it, to treat the foreign cells of the transplant as "self.") For example, in juvenile (i.e., type 1) diabetes, caused by a patient's own immune system wiping out their insulin-producing islet cells, one could imagine installing PD-L1 molecules on transplanted donor islet cells, thereby protecting them after transplant (see Bluestone, Chapter 21). A third option, of course, is to block the off signal and potentially get a stronger anticancer response by T cells.

"So immunology can be applied in a negative or a positive way depending on what your disease target is. We thought of all those possibilities."

But what tipped Freeman off to the potential application of his discoveries was the source of the information. "The EST we found [for PD-L1] was from an ovarian tumor, so we thought of cancer right from the get-go." It was still an educated guess, because a tumor mass is made up of lots of supporting cell types—as well as infiltrating immune cells—that are all distinct from the cancer. "But when we made PD-L1 antibodies, we went back and looked at ovarian tumor cell lines and PD-L1 was on them," says Gordon. PD-L1 expression was also found in breast cancer cell lines, and in fact, "Our pathologists saw PD-L1 expression on a lot of solid tumor types."

Clinical Trials, Here We Come

That would've seemed an obvious next step. It was, alas, not to be, at least not right away. Indeed, what happened next was more like a Dark Night.

Concurrent with the discoveries of Freeman, Honjo, and others was the 2001 approval of Gleevec, one of the most effective anticancer agents ever developed, with the FDA approving the drug in the fastest time to that date for any cancer treatment. Gleevec, however, was not an immunotherapy; Gleevec targets a kinase enzyme, part of a family of similar kinase enzymes that, when mutated (i.e., when they've gone "rogue"), can spur the development of multiple cancer types.

"Gleevec was an incredible home run," Dr. Freeman freely admits, "And basically the higher level administrators and funding agencies said targeting kinases [there are scads of kinases] is going to be the way to cure cancer. So everything after that was really supporting kinase work."

The result of this shift in funding was that they gave a quarter of Freeman's lab space to someone else.

"Those were lean years, absolutely," says Freeman, but the work continued. "A lot of our work in that period was also showing how PD-1 was involved in chronic infections." One of his findings was that if you get a chronic infection, the immune system goes into compromise mode: T cells start to express a lot of PD-1, making them susceptible to off signals, yet what happens is a sort of low-level insurgency. "For instance, if you have hepatitis it won't burn your liver out, but the immune system doesn't successfully eliminate the entire virus either." This phenomenon is seen in other infections as well: AIDS, malaria, tuberculosis—all chronic infections, and all involve PD-1. (For more on this see Schreiber, Chapter 7.)

"During that period a lot of our funding came through for the study of chronic infections, because immunology wasn't yet back."

What brought it back just a few years later was ipi. Shortly thereafter, the same company that pioneered ipi, Medarex, took on the task of developing anti-PD-1, which ultimately was named nivolumab.

"I think it's impressive that the same company that developed ipi developed the PD-1 [drug]," says Freeman. "That was a very astute group, and there I would give Alan Korman and Nils Lonberg a lot of credit." (Both researchers are now at Bristol-Myers Squibb.)

Anti-PD-1: Believe the Hype

Yes, Gleevec was an epic success, and yes, there are numerous effective kinase-targeting drugs now on the market, but the clinical trial results that eventually came out for anti-PD-1 pretty much blew the lid off the joint (see Topalian, Chapter 6).

Nivolumab was first approved for use in metastatic melanoma in July 2014, and other indications rapidly followed. "It's approved now for lung, kidney, and melanoma, and that's about 25%–30% of all cancer deaths," says Freeman, "And it's under FDA review for head and neck cancer, bladder cancer, triple-negative breast cancer," and probably others. Indeed, shortly after this conversation, nivolumab was approved to treat patients with Hodgkin's lymphoma.

The data are beyond impressive, but in some ways, the cancer-drug development world has seen this before, most memorably with a drug called Avastin, which was initially perceived as a nearly catchall cancer cure. Lo and behold, after reconsideration, it wasn't. Given the almost daily news reports coming out for immunotherapy, could this drug, this approach, be overplayed as well?

Freeman does have his concerns. After all, this is all very new. What about long-term side effects? "I worried about some late-appearing autoimmunity, something happening 10 years down the line maybe." But there are patients treated with ipi who are now 10 years out and more, and those sorts of side effects are simply not occurring. All the while, Freeman remains astonished at the drug's efficacy.

If only it had come along a bit sooner. "My mother had lung cancer and she died in nine months. She had the standard therapies, and radiation, and it just went very quickly." In the telling, Dr. Freeman's regret is palpable. "At the Farber they introduced me to some patients. There was a lung-cancer patient I met who was almost in hospice, who now rides his bike in for his checkups. He's fine," says Freeman, adding after a pause that, in fact, the only potential medical problem this once-incurable cancer patient has is that he's getting fat.

> *"There was a lung-cancer patient I met who was almost in hospice, who now rides his bike in for his checkups. He's fine."*

Of course, anti-PD-1 is not a cure-all. As with ipi, only a minority of patients responds to treatment: no more than 30% (except for patients with Hodgkin's lymphoma, where response is as high as 65%). Yet Freeman, who has the demeanor of a kindly guidance counselor or a priest, has not a doubt that these successes can be expanded on rapidly with new immunotherapies, or combinations of immunotherapies with standard therapies already available. "Absolutely!" says he, "I am a rabid advocate."

Freeman's advocacy extends to his ongoing investigations into yet another checkpoint molecule with the delightfully friendly name of Tim-3.

❧ ❧ ❧

Collaborators

Without exception, all the investigators profiled in this book stress the contributions of their collaborators, with the underlying message that science is not something you can do alone. You loners out there looking for career advice? Try computer programming. The biological sciences are not for you.

Beyond the collaborators already mentioned, Dr. Freeman emphasizes the contributions of Leiping Chen, M.D., Ph.D. (Yale), who firmly established the PD-1/PD-L1 pathway as a target for cancer, as well as helped organize the first-in-man clinical trial of the anti-PD-1 antibody, nivolumab.

Freeman also tips his hat to the contributions of Arlene Sharpe, M.D., Ph.D. (Harvard). "Her work was critically important in doing the biological assays that showed that cells transfected with PD-L1 [and sister molecule PD-L2] when used in this elegant mouse system with a T-cell receptor transgenic mouse, that those cells would specifically turn off." In other words, Dr. Sharpe nailed down the biological function of this molecular pair.

Speaking of pairs, in addition to being Freeman's frequent collaborator, Dr. Sharpe is Dr. Freeman's wife.

"We met as undergraduates at Harvard," says Freeman, eyes asparkle. "We both got our Ph.D. degrees in the Department of Microbiology and Molecular Genetics at Harvard Medical School, in different labs."

Thank you, Harvard.

The key to their happy relationship? One reason might be proximity. "We never shared the same lab. She's two blocks down the street." Also, there isn't much opportunity for competition. "We don't do exactly the same things. At that point in time in the early 2000s, I did molecular immunology, identifying genes and figuring out how they worked. Arlene did much more mouse immunology and mouse knockout biology. So we had relatively complementary skill sets."

And they don't take their work home.

"Well, when we do, my daughter yells at us. She's against going into the family business and is in international relations," says Freeman. "No, surprisingly, we talk about our kids, and how our days went. Personal things . . ."

The science can wait 'til morning.

CHAPTER SIX

Suzanne L. Topalian, M.D.

Professor, Surgery and Oncology
Director, Melanoma Program, Kimmel Cancer Center
Johns Hopkins University
Baltimore, Maryland

"It just clicked with me. It seemed to make sense." —S. Topalian

Suzanne Topalian was born in 1954 in Teaneck, New Jersey.

As with most, if not all of the investigators highlighted in this book, Topalian's attraction to science is intrinsic, but in her case not precisely a slam dunk. "I was always interested in science, but I was also interested in writing, so when I went to college I actually majored in English but I was pre-med," and for a time there was a bit of a tug-of-war. "My teachers and mentors in both of those areas were very encouraging," Topalian recalls, "And in the 1970s when I was in college, it was actually cool to be in medical school but to have a nonscientific background."

It's hard to imagine now, when higher education is often having your nose stuck to the single-discipline grindstone, "But in the 1970s people with a diverse background, that was accepted. Sure, there were people who knew more science and who had spent more time in labs during the summer who were in my medical school classes, but there were also people like me."

Ultimately, science won out. "But those writing skills are still very useful," says Topalian, obviously proud that she writes all her own papers.

Note: The use of ghostwriters in clinical science is fairly common, especially for clinical work sponsored by large pharmaceutical companies. Until fairly recently, this practice was covert, but now if an outside writer is used it's typically disclosed in the fine print.

Dr. Topalian graduated from Wellesley College and received a medical degree from Tufts University in 1979. Thereafter, a surgeon was born, courtesy of Thomas Jefferson University Hospital in Philadelphia. "This was a general surgical residency, but there was a major turning point during that time period when I took a year to do laboratory work at the Children's Hospital, which is right next door." Up to that point, Topalian was strongly considering a career in pediatric surgery but providence intervened. "The laboratory was specializing in tumor immunology," says she, "And that's where I caught the bug. I became very interested in cancer immunology."

A Road Less Traveled

At that point in her training, Topalian could have easily put out her shingle. She was already a skilled surgeon, so why switch gears?

"I had opportunities to go into practice after I finished my residency; I could have done that," and that work certainly had its appeal. "I went into it liking the idea of surgery because you get answers quickly." This is a personality trait one senses with her. Topalian would prefer that you cut to the chase, so to speak. "I mean, when you take a patient to the operating room, you think you know what's wrong with them and how you're going to fix it. Hopefully, it turns out that you were correct; sometimes there are surprises. But I liked that idea that you would pretty quickly know if you were right or not."

After a time, however, Topalian came to feel a disconnect both in the scale and the scope of her efforts. "Yes, it was gratifying to be able to solve problems and help patients ... one case at a time," she explains, "But it just seemed to me that there was a better way to have an impact on disease." The way to do that was through research.

> "...it was gratifying to be able to solve problems and help patients, one case at a time... But it just seemed to me that there was a better way to have an impact on disease."

Topalian was already in the right place to do that research, in training at Children's Hospital of Philadelphia in the laboratory of Moritz Ziegler. Ziegler was a prestigious cancer immunologist who was himself trained by the former Surgeon General of the United States, C. Everett Koop. (In science circles, your training pedigree is important. Your choice of mentor, and that person choosing you, says volumes about the quality of your training.)

"I did cancer immunology experiments for a year, and I really became hooked on the idea that the immune system could recognize and reject cancer the same way that it recognizes microbes and transplanted organs, etc.," says Topalian, "And it just clicked with me. It seemed to make sense." This realization dictated her next move: subspecialty training in surgical oncology, where the study of cancer immunology was part of the mission. Fulfilling these two criteria was a lab at the National Institutes of Health (NIH) run by one of the giants of cancer immunotherapy: Steve Rosenberg (see Chapter 13).

And there, Topalian thrived.

"He was a great ... I mean, we still have a great relationship, but he was a terrific mentor for me and for all the fellows. And I learned a tremendous amount, especially about how to design and conduct clinical trials," and the NIH is one of the very best institutions at which to do that. "It's a very special place," says Topalian. "You can't even be a patient there unless you're on a clinical trial. You'll never find that in any other institution." The place is so special that Topalian stayed on past the time of her fellowship, 21 years in all.

Aim High and Expect Nothing, or Onward through the Fog

Fellowships are generally the passing on of information and/or techniques that are already more or less known (with some room for refinement and discovery), but a fellowship in cancer immunotherapy in the early 1980s was, in many ways, starting from scratch. "I remember the very first project that I did with Steve when I walked in the door as a fellow was figuring out how to grow human T cells on a large scale so that we could treat patients with them." The technique was largely unknown, and those that were hadn't been applied this way before, yet the innovations required to make it work—just to grow the T cells—would be the underpinning of a major aspect of cancer immunotherapy, an approach called adoptive cell transfer (see Greenberg, Chapter 12). "So that was my job as a first-year fellow."

The project wasn't entirely de novo. The lab was already working with a T-cell growth factor called recombinant IL-2 (other growth factors remained to be discovered) and there was in-house experience in growing so-called LAK cells, a type of lymphocyte.

Lymphocytes and LAK cells: Lymphocytes are a subset of white blood cells, or leukocytes. T cells, B cells, and natural killer (NK) cells are all lymphocytes. LAK cells are NK cells that have been activated by the cytokine, IL-2, rendering them more cytotoxic than regular NK cells.

Cytokines are protein-based signals that instruct cells to do things. The signal can relay any number of commands, ranging anywhere from "come hither" to "go forth and multiply" to "seek and destroy."

Linda Muul, the person in charge of the LAK cell effort in the Rosenberg lab, taught Topalian what was already known. In brief, take some peripheral blood from the patient, separate out the cells you want, feed the cells some IL-2, let them sit in nutrient culture for a few days, and then give them back to the patient. It was all rather straightforward.

However, explains Topalian, it is not so straightforward to take a solid tumor out of the operating room, extract lymphocytes from it, grow the lymphocytes in the right way so that they'll have tumor-specific recognition killing and cytokine secretions, verify that the cells can actually recognize the patient's tumor, and then give them back to the patient. "And that culture could go on for a few weeks," says Topalian, "So that's actually not such an easy thing to do and that's what we had to develop." (See Rosenberg, Chapter 13; Hwu, Chapter 15.)

It was a formidable task, but any discouraging taunts from initial out-of-the-box failures were kept at bay with humility and plain old hard work. "There were a lot of things that didn't work," says Topalian. "I had actually never cultured a human lymphocyte before I joined Steve's lab, so there were a lot of things that I had to learn." But she was determined, and she had the drive. She read. She consulted. She read some more. "I actually spent a lot of time talking to technicians who were doing these things day in and day out, and these people actually welcomed me at the research bench, and I'll always remember that." That personal touch plus the hands-on technique were invaluable. "One of the senior technicians made a little space for me right next to his own bench and I was shadowing him," and that's about as proactive as you can get. "I've picked up as needed to do what I needed to do," Topalian says, which is not exactly a sink-or-swim approach, but learning what you need when you need it to stay afloat in the moment.

This approach is predicated on doing your homework. Yes, much of what the NIH lab was doing was high-risk research with a multitude of unknowns, but with the proper preparation and an informed approach, you can stack the deck in your favor. "I've always tried to maintain focus and try to do as much groundwork as possible before I launch into a project," says Topalian, "But having said that, they haven't all worked out."

Among the things that didn't work out in the long run? A therapeutic approach using LAK cells.

After her four-year fellowship came to an end, Topalian was given the title of Senior Investigator, the equivalent of a faculty member at the NIH, and her own laboratory.

PD-1:
Up, Up, and Away

PD-1 showed up on Topalian's radar around 2003, and although novel in the specifics, the concept was by that time familiar to her from having been a co-investigator on the ipilimumab trials. "We wrote the first few papers not only on the efficacy of anti-CTLA-4 but also on the characteristic side effects, the immune-related side effects. That experience was extremely important in developing anti-PD-1 and anti-PD-L1," says Topalian. "It really accelerated the process."

Her experience with ipi was not merely instructive; it was motivational. "With ipilimumab, the first thing we actually saw before we saw a tumor regression was ... immune-related toxicities," and that was a good thing. As any drug developer will tell you, a drug that is utterly without side effects is a drug that doesn't work. "And everybody was, you know, 'Hurrah!'" Topalian recalls. "I think that was the first time in my career that we were happy to see a side effect because it meant the drug, mechanistically, was working the way we thought it would."

"I think that was the first time in my career that we were happy to see a side effect because it meant the drug, mechanistically, was working the way we thought it would."

The observed side effects were also a selling point. The adverse events for ipilimumab could be fearsome, but less so with anti-PD-1.

"Now that many, many thousands of patients have been treated with ipilimumab and tremelimumab, the [anti-CTLA-4] Pfizer drug, we know that the side effects to benefit ratio is not ideal there," Topalian explains. Yes, the long-term survival for desperately ill melanoma patients was unprecedented—around 20%—but the rate of serious adverse events hovered around 20% as well, which is high. "So we knew we had to keep looking, and it was known on a basic scientific level that there were other immune regulatory pathways," and there in front of her was PD-1, and even at first glance it just seemed like a better idea.

What Topalian knew at the time was that PD-L1 was the major ligand (i.e., the thing that turns on the "don't shoot" signal) of PD-1, but that unlike

CTLA-4, PD-1 was expressed on human cancers and not on most normal tissues. Further, working in collaboration with Lieping Chen (of Yale), it was shown that PD-L1 is expressed in many human cancers. That really got Topalian's attention. "At the time Lieping [then at Hopkins] and Drew [Pardoll], who was here at Hopkins since forever, started talking to me about PD-1 because my specialty is studying human tissues in in vitro systems. I don't work with murine models, but they work with murine models, so they talked to me about the importance of this pathway and I started experimenting with it while I was still at NCI [the National Cancer Institute, part of the NIH]."

It didn't hurt that the environment of the NCI, by its very nature, is protective of investigators with daring ideas. "Let's face it, I was in a specialized environment in the NCI. Everybody I saw everyday was engaged in the same kind of work," says Topalian. "And we were seeing responses—not at the level that we wanted to—but we were seeing some patients respond to each of these various forms of immunotherapy, enough that the work kept going."

It also didn't hurt that, once again thanks to ipi, Topalian already knew what tumor type to test with this new anti-PD-1 agent. "Melanoma has always been the poster child, the case study for immunotherapy," explains Topalian. "We figured out from the days of IL-2 that, for whatever reason, melanoma responds to immunotherapies of any kind at a higher rate than other solid tumors. Kidney cancer is number two. Nobody to this day really knows why."

Clinical investigations in these two tumor types were successful enough to earn rapid approvals for anti-PD-1 and served as launch pad for exploring its activity in other tumor types.

"Now we're all excited about [anti-PD-1 and] Hodgkin's disease, and triple-negative breast cancer, and squamous head and neck cancer, and gastric cancer, and hepatocellular ... I mean, there's now a growing list of cancers that we're excited about."

> "Now we're all excited about [anti-PD-1 and] Hodgkin's disease, and triple-negative breast cancer, and squamous head and neck cancer, and gastric cancer, and hepatocellular . . . I mean, there's now a growing list of cancers that we're excited about."

Moving Forward, with Caution

Sad to say, anti-PD-1 is not for everyone. "We know that prostate cancer so far has not responded to these drugs when you give them as a monotherapy," says Topalian. "We also know that most patients with colon cancer do not respond ... , but on the other hand there are other tumor types that haven't been tested yet, and the fact that so many tumor types are now responding to anti-PD-1, it at least raises the possibility that there are additional tumor types that might respond. I wouldn't call that hype, but I think that's based on a reality." The bottom line for the moment is that there is a growing list of cancers that have a finite response rate to anti-PD-1, and these responses are in patients with very advanced disease who have not responded to other therapies before. "We're seeing cancer regressions in difficult-to-treat patient populations, so there's reason to be optimistic about that."

Two ways to maintain that optimism are (1) the identification of biological markers (like in a blood test) that would help determine who would most benefit from treatment, and (2) the potential of combination treatments with other drugs. "And those two issues are linked," Topalian says. "We may very well be able to find markers in the tumor microenvironment that would indicate bypass pathways or resistance pathways that could be co-targeted with other drugs, so this is what we're focusing on right now at Hopkins."

Unfortunately, that focus can cost a fortune. "Industry is an extremely important player in everything we're doing now. They are a very important source of funding. These kinds of clinical trials, for the most part, they can't be funded or even conducted at a single institution." But, those dollars are still nowhere near enough. Topalian often finds herself having to financially justify her investigations.

"It has become very difficult to fund science; it's now necessary to be very entrepreneurial in terms of finding opportunities in all kinds of places," she warns. "The NIH has not been a reliable source of funding for the past several years." Not only are the award rates very low, but even after you get the award, the funding can be cut back the following year. "It's not based on performance or anything, but just simply an adjustment given the amount available."

This has forced Topalian to be creative. "We have grants from the NIH, from industry, from foundations—disease-based nonprofits have become a very important component of funding now—and philanthropy," Topalian says. "If you add up all of that in terms of what we have to do to connect with those sources of funding and compete for funding, I would say it takes up easily a third of my time."

The Right Fit

Eternal grant writing may not be your cup of tea. Working 12- to 14-hour days, and often weekends, might not float your boat. Spending a lifetime to painstakingly elucidate something that no one ever really wanted to know (until you convinced them they should) is not for everyone. But if you're going into science, that's one of the first research questions you should ask: Is this right for me?

"I don't think I've ever told anybody in those terms that they were not suited," says Topalian, who has trained many young scientists. "But it usually becomes clear, I mean, not only to me but also to that person, that the demands of the job and what they want in their professional life don't match." However, reaching that decision need not be a disastrous all--or-nothing choice. "Science is a very diverse profession, and so you can contribute at all levels." Maybe the best choice for a particular person is not going to be at the bench thinking of experiments. Maybe that person would be better in science writing, science regulation, or science law. "There are so many possibilities," says Topalian, ever forward-thinking. "So I help people try to find the good fit. I have never discouraged anybody from being involved in science."

Or, for that matter, being involved with scientists.

Dr. Suzanne Topalian is married to Dr. Drew Pardoll (see Chapter 8).

Oddly, even those who have known one or the other of them professionally for years do not know this.

"Yeah, there are several people out there that don't know that." Topalian shrugs, matter-of-factly, "I don't know … I don't think either one of us is hiding it. We've been married for 22 years." The marriage is also a research collaboration. In fact, if you see Suzanne present her work you will notice, if you look closely, that Dr. Pardoll is credited on any number of her background slides. Should you see Drew present, Dr. Topalian's work is often cited.

All in the name of science.

⚜ ⚜ ⚜

"While I was in medical school, my father, who was not a scientist or a physician, asked me how things were going in medical school and specifically he wanted to know why it was taking so long to cure cancer, and this was 1975, four years after the war on cancer was launched. I patiently explained that cancer was really at least a hundred different diseases and that this was a very complex situation. But his response was, 'This is not complicated, it should be simple, all you need to do is find the common denominator.'"

And it turns out that Mr. Topalian was right—if not just a bit psychic—because his daughter, 20 years later, was responsible for the seminal paper "Immune checkpoint blockade: A common denominator approach to cancer therapy" (Topalian S, et al. *Cancer Cell* 27: 450 [2015]).

SECTION III

IMMUNE SURVEILLANCE

Cancer Immunoediting
9/28/16

Innate + adaptive immunity

| Elimination | | Equilibrium |

Treg

Treg

MDSC

PD-1 / PD-L1
CTLA-4
LAG3
TIM3
VISTA

MDSC

Escape

Immunosuppressive

Microenvironment

| Escape |

ROBERT SCHREIBER
"Cancer Immunoediting"

CHAPTER SEVEN

Robert Schreiber, Ph.D.

Director, Center for Human Immunology
and Immunotherapy Programs
Washington University School of Medicine
St. Louis, Missouri

CANCER IMMUNOEDITING

"We all know that cancer immunosurveillance doesn't exist."
—R. SCHREIBER, QUOTING A REVIEWER OF ONE OF HIS EARLY PAPERS

Cancer immune surveillance *does* exist and the man who proved it was Robert Schreiber, born in Rochester, New York in 1946.

Dr. Schreiber obtained his undergraduate and graduate degrees at the State University of New York at Buffalo. He completed his postdoc education at the Scripps Research Institute, located in La Jolla, California, home to the lush La Jolla cove, the crown jewel of the La Jolla Underwater Park Ecological Reserve.

The protected cove is a premier site for scuba diving. Schreiber does not dive. Schreiber swims, like on the great TV show (or the bad movie) *Baywatch*.

"I rescued a dolphin once," Schreiber says, beaming (he beams a lot). "When I was at Scripps, it was still located in downtown La Jolla so if you didn't get there early in the morning you'd have to park, well, who knows where." One day he was so late he had to park far north of Scripps, which meant walking back along the coastline. "As I was walking, I saw this dolphin that had washed up on the shore, and it was still alive. And so I came down and jumped in the water and kept it wet." As Schreiber ministered to his charge he alerted a passerby to the situation, who then called SeaWorld, who came and took the dolphin away. "I ended up having to go all the way back home to change because I was soaking wet, but it was a great experience."

After his fellowship at Scripps, Schreiber joined the faculty of Washington University in St. Louis in 1985 as Professor of Pathology, where he works to this day.

Prior to his current position, there were the mentors, because there have to be; you don't get to this level of achievement on your own.

"My dad was a chemist; he worked at Kodak and was actually one of the developers of microfilm," says Schreiber. "He was an organic chemist by training, and so I was always fascinated by that." Deciding what he wanted to be when he grew up was just a matter of what sort of scientist he would become: physician or bench scientist? "When the time came, I decided to do research because I worked in a laboratory in college through my junior and senior years chopping up brains and isolating sodium- and potassium-stimulated ATPases [enzymes]." Cow brains, actually. He was chopping up the brains of cows, which are quite large, while looking for something really quite small. This is the work of a scientist.

"I really enjoyed the work," says Schreiber. Also, chopping up brains still served the purpose of being an exercise in chemistry—albeit biochemistry, with its intricacies of living systems and potential medical applications. Fascinating stuff, to be sure, but it had nothing to do with the immune system. Schreiber's focus on immunology came entirely, and quite literally, by accident.

"There was one amazing sort of turn in my career," says Schreiber, and it involved the biochemist, Professor Jim Watson (no relation to the James Watson of DNA double-helix fame). "I was planning on staying in Buffalo and doing my thesis work with him, but he was killed in an automobile accident, and suddenly my entire future just fell apart."

Fortunately, Dr. Watson had a best friend who was an immunologist, and this immunologist friend had just opened his own lab and needed graduate students to fill it. "He said, 'If you've ever thought about immunology and were to stay and do your graduate work here, I would be happy to take you into my lab.'" That would be the lab of Dr. Morris Reichlin.

"He was a rheumatologist studying antibodies against proteins, which nobody did at the time." (In this context, the word "against" means "matched to" or "tailored for." Antibodies cling to what they were raised "against.") "He had a beautiful system where he compared hemoglobins," explains Schreiber. The system was able to discern the blood protein hemoglobin A, which is the normal hemoglobin, from hemoglobin S, which is the dysfunctional hemoglobin from sickle cell anemia. The two proteins differ by only a single amino acid (the building blocks of proteins) so raising an antibody against such a subtle difference was a neat trick.

Yet, nifty tricks aside, why immunology? The vocabulary alone can give you an aneurysm.

"To me, immunology was the perfect fit, my connection between medicine and chemistry," says Schreiber. And you don't have to explain to others outside the field about why you were doing what you were doing because you are attempting to understand disease susceptibility: how to protect against disease, how the immune system can actually be harmful instead of helpful. Who could argue with that?

> "To me immunology was the perfect fit, my connection between medicine and chemistry."

Fascinating, but by no means was Schreiber thinking about curing cancer.

"Cancer was the farthest thing from my mind at the time. I worked in the area of complement."

> **Complement** refers to a plethora of small proteins that circulate in the blood to help antibodies and immune cells clear invading microbes and cellular debris. They do this in one of several ways: by latching on to the target and acting as a beacon for other components of the immune system, by rounding out the attack of an antibody that has already latched on to an antigen, or by attacking the invader directly en masse. Complement is part of the innate (i.e., unchanging) immune system and is ancient in origin, going back hundreds of millions of years—even before the appearance of vertebrates.

"I studied the proteins in the complement system," says Schreiber, "and I was the first person to purify the fourth component of complement C4, an unusual protein. It had three polypeptide chains and that was very unusual."

This work, with the results described so briskly above, took more than a year to perform. As Schreiber points out, with a sort of "joke's-on-me" chuckle, is that he made his discoveries before the advent of gene sequencing and any number of other modern technologies that would have aided in the task.

What Schreiber did was grunt work. To study a protein's structure was to purify a pile of it and try to tease out the precise nature of the structure by observing what happens when it became active (proteins, like most mechanical devices change shape in some way when activated).

The first step of this process was to separate the desired complement protein from all the other proteins found in a sample of blood, a large sample—in this case, enough to fill a vat. "Oh, my God! We had purification columns [glass or plastic tubes] filling a space as big as this room! [That room being the medium-sized Starbucks where this interview took place.] Buckets and buckets and buckets of blood, and you would put them on these columns and you'd put them in the cold room and you would watch these things run down the columns."

As a sample runs down a purification column, proteins physically separate according to their size (just like the small pieces of granola finding their way to the bottom of the cereal box). At the end of the purification column is a nozzle that drips out uniformly sized drops at some set rate to be collected in small vials by an automatically rotating fraction collector; thus, drip, drip, collect, turn to next vial, drip, drip, collect, next vial, and so on. All night, in the walk-in refrigerated cold room, which is where you have to do experiments on materials that will rot if not kept cold.

Ideally, the scientist loads up a column in the cold room and walks away. Hours later, after a good night's sleep in a warm bed you retrieve the samples. "But the fraction collectors were only partially accurate," says Schreiber. "You just knew that when your protein was coming off the column the fraction collector would jam," and your purified sample would wind up on the floor of the cold room. "So you would sit in the cold room all night long and babysit the column. I can't tell you how many frozen fingers I got out of that project."

Still, it was a thrill. "It was so exciting to see, starting with this incredibly heterogeneous mixture, which was blood, and ending up with a single protein that on a [separation] gel was a single band, until you reduced it and then you saw the three bands."

Voila! Complement C4.

Schreiber went on to elucidate a particular aspect of complement referred to as the alternative pathway, and then shifted focus to what would be the foundation for the rest of his life's work, **adaptive immunity**.

---

*The immune system is comprised of two parts: the **innate**, and the **adaptive**, immune systems. The **innate** system is easier to understand: It is the police force on patrol looking out for generally suspicious characters in whatever common form they might take — in a word, profiling. The innate immune system includes the skin, mucus, saliva, and sweat, which all work to prevent pathogens from breaking in, or to destroy them if they do. It also includes certain white blood cells such as natural killer (NK) cells, mast cells, the phagocytic cells, and more. They are innate because they look for harmful activity, but they do not deliberately target or remember specific pathogens; they "know what they know," but don't learn and thus don't remember what they've seen.*

*The **adaptive** immune system, on the other hand, does target and it does remember. The adaptive system is composed of B and T cells, each of which is like a detective on the hunt for a single antigen, a single criminal. These proteins, or antigens, are markers that B and T cells can recognize as belonging to harmful agents. Viruses, bacteria, parasites, and cancer cells all have antigens. If adaptive cells ever run into their target antigen, they will remember it and try to destroy it the next time it comes along. Both the adaptive and the innate immune systems play a role in cancer, but* *tumor immunologists are especially interested in the adaptive immune system for the possibility that it could remember the cancer, and automatically destroy it were it ever to return.*

Crucial to that "remembering" is the ability to discriminate: Adaptive immune cells can distinguish tumor cells from healthy cells, something that chemotherapy, radiation therapy, and even most "targeted" therapies cannot.

Then Came Cancer

"A lot of my career was guided by finding the right mentors," says Schreiber. For the complement work there was the German-born Hans Müller-Eberhard, who ran a large laboratory at Scripps. Then there was Cuban-born immunologist, Emil Unanue (then at Scripps, now at Washington University), who introduced Schreiber to macrophages, a type of innate immune cell that engulfs its target whole, like an amoeba.

"My transition into cancer really came many years later, not until about 1988 or so when I met Lloyd Old," says Schreiber, thinking back, taking stock. "That name will probably appear in many of your stories along the way. Incredible man…, brilliant and forward-thinking, and in many cases actually influenced people into doing cancer research, into what is now called immuno-oncology, when they had no intention of ever doing that. That's how I got into it."

Storm Clouds on the Horizon

"You can't do science if you don't fail," says Schreiber, again beaming, perhaps thinking of some glorious benchtop flop. ("Benchtop" refers to a scientist's work space, and a type of scientist. Physicists, whose work is theoretical, do not require benchtops.) "I think that's what distinguishes the people who succeed in science and the people who don't." Yes, you have to love what you do, and yes, you need to be driven, but perversely, to be successful, the focus of your work should be to disprove your hypothesis rather than proving it. "It's that chase for new knowledge that drives most of us." If that chase proves you wrong, so be it.

And there has to be some fight in you. Confronting the unknown has never been for the timid.

It's like this: You come up with an idea. You have data that supports the idea, but there are people that say no, you're wrong, because they've done their own experiments and have arrived at the opposite conclusion. "When I was in complement we found this three-chain structure of C4, but another group thought that it was two chains," says Schreiber. So, you listen politely, and then you defend—not outwardly at first, but within. You repeat the experiment. You do it again, and again, and again. "And every time you do it you find that it's three chains, and then you realize, you know what? I'm right, they're wrong. But that period of time where the idea is under attack is the time when you say, I've really got to do everything I can to make sure that my result is correct." Let the data be your shield.

That was Schreiber's experience with complement. The experience with immunology and cancer and something called "immune surveillance" was fundamentally different; during the complement work, no one ever made the assertion that complement did not exist. "But when I started working on [cancer], 99 out of 100 people would tell you cancer immune surveillance doesn't occur, that cancer cells are too much like normal cells and that the immune system can't see cancer because there's no danger signal in cancer." No alarm bells, no signal, no tip-off—as there would be from a virus or microbe—that something in your body was amiss at the microscopic level.

Yet, the data from Schreiber's animal experiments were telling him otherwise. "And for five years we were getting beautiful data. We would submit the data and the reviews would come back saying, you're wrong. In fact, the direct quote from the reviewer was, 'We all know that cancer immunosurveillance doesn't exist.' That's the kind of thing that I faced," says Schreiber. "But that was the aha moment: when the reviewer (with whom Schreiber is now good friends) said there is no such thing, we'd done the experiment. *We had the evidence.*"

And then came Stutman. Osias Stutman.

"So … the Stutman experiment." Schreiber pauses. He takes a deep breath. "They were actually very well-done experiments. The problem was the timing." Stutman's work seemingly disproved Schreiber's hypothesis by demonstrating that the immune system was completely blind to cancer. To arrive at this conclusion, Stutman used a special type of mouse bred to have little, if any, immune system to fight off assaults of any kind, be it the common cold or cancer. Stutman injected these mice, as well as a batch of normal mice (the control group), with a cancer-causing agent. The experiment showed that both types of mice developed cancer at the same rate, suggesting that the presence or absence of an intact immune system has no effect on the development of cancer.

Pretty straightforward, really, so what was the issue with the timing? "There was only one immunodeficient mouse to work with at the time," Schreiber explains, and that was the so-called nude mouse. (The mouse is nude because the genetic manipulations that erased its immune system also removed all its hair. The exact mouse strain referred to here is called CBAN.)

What Stutman didn't know at the time of his experiment was that, in fact, there are some T cells left in CBAN nude mice, just not that many. He also didn't know that there were still NK cells, or that the CBAN strain of mouse in which he did the experiments had the highest specific activity of the enzyme that converted methylcholanthrene (the cancer-causing agent used in the experiment) into its carcinogenic form. Thus, the CBAN mice developed more cancer when exposed to this agent than one would normally expect.

In aggregate, Stutman, unbeknownst to him, was trying to win an argument based on a critically flawed premise. No experiment, no matter how elegant, can overcome that. "So, when he put this carcinogen into CBAN mice versus normal mice he didn't see a difference in the development of the tumors between the two. And he said, okay, it's showing you that there's no such thing as immunosurveillance," says Schreiber, "And Stutman was absolutely correct in terms of his interpretation based on what he had in hand," But, at the end of the day it was a matter of "garbage in, garbage out."

Which says nothing about Dr. Stutman personally.

"He was quite a gentleman," recalls Schreiber. "I remember giving a talk at Memorial at a small meeting and he walks in and sits in the back of the room—mind you, I had never met him—and I said to someone sitting next to me—I think it was Ron Levy (of Stanford)—I said, 'Who's that person?' And Ron says, 'Oh, it's Osias Stutman.' And it was like a gunshot. So, I presented my results and Stutman's the first to raise his hand and I call on him and then I'm just grabbing onto the podium and I'm going, 'Oh my God, here it comes.' And he gets up and he says, 'You know, this is what you can do now [in the early 1990s] that I could not have done in 1971.' I thought that was the most amazing thing for someone to say. I so respect him for saying that." Dr. Stutman, after all, was a scientist through and through.

And yet, even with matters thought settled there will always be detractors.

"There's another investigator who's been kind of like the thorn in my side, who for some reason cannot reproduce our results despite the fact that our results have been reproduced in probably 30 laboratories all over the world."

That would be Dr. Thomas Blankenstein of the Max Delbrück Center for Molecular Medicine in Berlin.

"He's written commentaries about our work, coming up with all kinds of alternative explanations other than if it walks like a duck and quacks like a duck, it's not a horse," says Schreiber, not beaming, but still with a smile. "And so the interesting thing about Thomas—who's a very good scientist, he's very well respected—is that this friction between our results and his has brought a lot of attention to the field and as such, I think it actually helped."

Cancer Immunoediting

Rather than continuing to repeat experiments to quell the remaining skeptics, Schreiber decided to investigate anew. "We had two options: We could continue to do the same kind of experiments where we were comparing carcinogenesis in wild-type (normal) versus immunodeficient mice, or we could choose to move forward where, if this immunosurveillance is real, then the next logical prediction would be so and so, and then test that logical prediction. And that's where we chose to go."

Accepting that the immune system can "see" cancer, one of the first predictions was that tumors somehow protect themselves from the immune system not at the time of first attack, but after an initial immune response. This prediction was proven by using any number of differently immunocompromised mouse models.

"We predicted that there would be a phase in which most of the cancer would be killed off by the immune system," says Schreiber, "There would still be some cancer remaining, but the immune system would keep it in a state of dormancy," a molecular stalemate, if you will. Anecdotally, oncologists see this all the time where patients can have dormant or very slowly progressing cancers. "And we predicted [and then proved] that the immune system was a mediator of that dormancy."

The experimental proof of this idea resulted in a *Nature* paper. (Publishing in *Nature*, one of the most prestigious science journals in the world, is a big deal. No scientist will ever hesitate to tell you they had a paper in it.) "The first paper was a *Nature* paper, the second paper was a *Nature* paper," and when the concept of immunoediting was fully elucidated, that was a *Nature* paper too. "That's the paper that really made the difference, when we published the mechanism by which **cancer immunoediting** occurs."

> *Cancer immunoediting: All the cells in a tumor are genetic mutants of some sort that led them to become cancerous, but not all of the cells are mutated in exactly the same way. Some of the mutations are easier for the immune system to see than others. When the immune system first encounters the tumor, the T cells will kill off the most immunogenic (i.e., easily seen) cells, leaving behind tumor cells that are less vulnerable to attack. Following this initial attack, the surviving tumor cells send out a signal telling the immune system to stop the onslaught. That signal is called a checkpoint. (See Sections I and II.)*

"We were obviously helped by Jim's work," says Schreiber. ("Jim" in the IO world is like "Angelina" or "Brad" in the rest of the world—no last name needed.) "And in many cases our work helped him explain what was he was seeing." There also was independent confirmation of the cancer immunoediting hypothesis. "A tumor biologist, Tyler Jacks [of MIT], used a system which was different than ours and came to the same conclusion. So [our group] and a very well respected tumor biologist published back-to-back papers in *Nature* saying exactly the same thing. And there you go."

Bob Schreiber explained what was going on with the immune system. James Allison showed how to fix it. "That was our transition from going from real basic tumor immunology into what turns out to be cancer immunotherapy."

Caveat

To date, the survival gains for cancer patients achieved with cancer immunotherapy are without precedent; it's headline stuff, and deservedly so, but it is not the end of cancer. Is it the beginning of the end? Perhaps. But there's lots of work that remains to be done.

"I think the hype is dangerous."

"I think the hype is dangerous," says a beamless Schreiber, who has seen it all before. "Ten, 15 years ago we were hearing every other week that somebody has cured cancer. Eventually, it became obvious to people that we hadn't cured cancer and people were saying, what is all this crap that's being talked about? And so we became, as immunologists, the laughing stock of the tumor biology community." Claims were being made that simply had no validity to them that were not scientifically justifiable.

Have there been home runs with cancer immunotherapy? Yes, beyond doubt. However, as yet, the majority of patients don't respond to immunotherapy, and for those that do, long-term outcomes are still in question. "What happens if suddenly these patients who have been remaining tumor-free for 10 years after treatment with ipilimumab suddenly break through and there's actually cancer cells?"

Note: As of April of 2017 this has already happened: A melanoma patient relapsed after eight years of being supposedly cancer-free.

Carefree

Beaming anew—and inwardly, probably without end—Robert D. Schreiber says: "I love what I do and I love the people." The work keeps him riveted, and the people keep him young. "You know, I'm going to be 70 this year and I look around my lab and I'm dealing with 30- to 35-year olds on a day-to-day basis and it's fantastic."

He plays, too. "I love golf. I stink, but I love golf. And I also love the camaraderie. I play golf with two foursomes and most of them are scientists. We go out there and we laugh for four hours on the course, and that's great. Of course, I'm usually the brunt of the jokes, but nevertheless, it's a great time to just let go and do things like that. There's no talk of science necessarily, but it's just being with friends."

Curing cancer. A few laughs. What more could one ask for?

SECTION IV

VACCINES

DREW PARDOLL
"Bad Neighborhood"

CHAPTER EIGHT

Drew Pardoll, M.D., Ph.D.

Co-Director, Cancer Immunology
and Hematopoiesis Program
Professor of Oncology
Johns Hopkins University
Baltimore, Maryland

"Vaccines were failing right, left, and center."   —D. PARDOLL

Drew Pardoll was born in 1956 in Newark, New Jersey. "But I was raised in Elizabeth, which is half a notch better," he says with a laugh or, really, more of a giggle. Drew giggles. He seems perpetually thrilled about most everything. It's infectious.

What thrilled him as a boy was the idea of being a scientist. "That's all I was ever interested in. The only two birthday presents that I ever cared about getting in my life—in fact, they're the only two presents I ever remember getting—[were] a microscope and a telescope." He was nine. "The only question was whether I was going to be an astrophysicist or a biologist." Obviously, biologist won.

The scholastic progress toward that end was swift. As an undergraduate at Johns Hopkins, Pardoll got his first mentor and published his first paper. It was a significant achievement academically and, further, the process introduced Pardoll to the life lesson of fighting for what you believe in.

This particular fight was with a reviewer. As with any paper submitted to a reputable journal, this paper had to survive peer-review. In this case, one of the reviewers had been particularly harsh in his criticism. Worse, this reviewer was particularly famous. "Remember the Meselson–Stahl experiments?" asks Pardoll. (Everyone in biology remembers the Meselson–Stahl experiments; it's in all the textbooks. If not, see Meselson–Stahl in the Glossary.) "It's one of the most elegant experiments in biology." The harsh reviewer was none other than Franklin Stahl.

"He actually had a reputation for being really hard," but Stahl was one of the gatekeepers. Pardoll wanted in, and he did everything that was required to gain Stahl's approval. In the end, the science was solid and the paper accepted, although the victory was bittersweet. Just before the paper came out, Pardoll's mentor, Hillard Berger, died of complications related to type 1 diabetes.

"I remember being sort of devastated," says Pardoll, "But at the same time very proud because ultimately we were able to address the reviews and get it published." Crowning this pride came the moment when he got his physical copy of the article and actually held it. In the recollection, Pardoll holds his hands in front of him, palms up, supporting the invisible journal, "It was the coolest thing in the world. Just the coolest thing."

10 Eye Opening

Pardoll continued his training under the tutelage of Don Coffey and Bert Vogelstein (both still at Hopkins), and he continued to impress by publishing not one, but two papers in 1980 in the premier journal *Cell*. However, the experiments described in those papers had to do with the nuts and bolts of DNA replication, not immunology, and certainly not cancer immunotherapy. The correlation between cancer and an immune response came just a bit later while Pardoll was doing an oncology fellowship in the bone marrow transplant program at Hopkins. During that time, he was introduced to the awesome power of the immune system.

It was horrifying.

"That was before we had cyclosporine," says Pardoll, meaning that there was no way to effectively dampen a patient's immune response after receiving a transplant. "So folks got really bad **graft-versus-host disease** [GvHD]. I don't know how much you know about GvHD but it's really ugly. I mean, you can literally slough off your entire colon in one bowel movement . . . It was the hardest service in the world."

Graft-versus-host disease: *GvHD would be a smackdown in the middle of a wrestling match, if the wrestling match were between the graft or transplant (in this case, the bone marrow of a healthy donor) and the patient who receives the transplant (the host). GvHD occurs when the immune cells of the donor (the graft) perceive the entire host as foreign and tries to eradicate it. The host tissues fight back, trying to suppress the attack by the graft's immune cells. This tussling persists until a standoff is reached, and the newly transplanted immune system learns to tolerate the host, and vice versa. If not, then either the graft is lost or the patient is lost. This is why patients take immunosuppressive drugs after a transplant of any kind. Unfortunately, even with such precautions, more than 10% of transplant patients experience a serious, potentially life-threatening, bout of GvHD.*

There was a bright spot or two during that time for Pardoll, a few moments when the situation's "gee-whiz" factor outshouted its nightmarish downside. "I

> "I used to call it Spaceship Onc because the techniques really were space-age."

used to call it Spaceship Onc because the techniques really were space-age," says Pardoll, describing how the transplant team first ablates (wipes out) the patient's existing, cancer-ridden immune system and replaces it with the immune system's precursor: the bone marrow of another person, often a complete stranger. If all goes well, the patient gets to live. "What we used to tell patients was that if they survived their graft-versus-host disease, their leukemia would never come back."

This was space-age technology to be sure, but like rocket ships, these transplants were subject to intrinsic disasters. "To keep my sanity I would get copies of *Science, Nature,* and *Cell* shipped to my house. On the occasional times that I actually went home, I'd grab them and bring them back with me to the hospital and around, say, 11 o'clock at night when I had a few free moments I'd flip through."

On one such bleary-eyed night, Pardoll flipped through, but then reread very carefully two back-to-back papers by Mark Davis (of Stanford) and Tak Mak (of the Princess Margaret Cancer Center; see Chapter 11) on the cloning of the genes for the T-cell receptor. This work was foundational in the field of immunology, providing the basic molecular understanding of how T cells work, and how they might be clinically deployed.

"And here I am not knowing anything about immunology but I'm witnessing the power of the immune system every day in the bone marrow transplant program and I'm reading this article and BOOM! It was instantaneous," explains Pardoll as his smile returns. "So, yeah . . . I can be smarter than graft-versus-host disease. I can use molecular biology to figure out how to target the immune system much more effectively against the tumor."

It was a dual epiphany. First, Pardoll realized he needed to know everything there was to know about T cells. Second, given the faculty at the time, he could not learn those things at Hopkins. Pardoll left to seek further training at the National Institutes of Health (NIH) where the study of T-cell immunology was already well underway.

"It took me about 10 minutes of interviewing at the NIH with Steve Rosenberg (see Chapter 13) to realize that there was no way I was going to do my training there," Pardoll recalls. Rosenberg was knee-deep in the field of cancer immunotherapy, but for Pardoll's needs, the NIH work put the cart before the horse. Rosenberg was not interested in the mechanistic underpinnings of how the immune system worked so much as he was interested in using the immune system to cure cancer. "Don't get me wrong, I have huge respect for him. He's a really impressive individual," but Pardoll's philosophy dictated that he needed to fully grasp the fundamental reins of basic immunology before treating patients.

"So I studied T-cell development with a guy named Ron Schwartz, who's a very good immunologist in the NIAID [National Institute of Allergy and Infectious Diseases]." The work was all basic immunology, no cancer. The cancer investigations began after Pardoll returned to Hopkins.

Vax Attacks

At the time Pardoll began his foray into cancer immunology, patients were routinely receiving cytokine-based drugs experimentally in the hopes of giving their immune systems a boost. However, with the single exception of melanoma patients treated with the cytokine IL-2 (where overall responses were still just so-so), all other cytokines tested were nonstarters. To Pardoll, this was no surprise.

"Having been trained as an immunologist, I learned that one of the fundamental understandings for cytokine and lymphokine biology is that they're meant to work in an autocrine or paracrine fashion." Meaning, they are not like hormones, which work systemically; cytokines work within a few cell diameters of the cell producing them. "So it was a completely nonphysiologic approach to try to use them by giving these massive doses [systemically] like a drug," says Pardoll. It made way more sense to deliver the drug at the site of therapeutic need. "So I had the idea to put the IL-2 gene into tumors . . ."

This work, performed in mice, was, as they say, "hypothesis generating." That means the experiment was not directly successful so much as it suggested other experiments that might be. The hints as to what to do next took Pardoll to the lab of expert gene jockey, Richard Mulligan (then at the Whitehead Institute, and now at Harvard), who proposed a systematic approach to find the best cytokine for the task at hand.

Working in Mulligan's lab at the time was an oncology fellow by the name of Glenn Dranoff who was exploring the nexus between gene therapy and cancer treatment. Dranoff's project in the Mulligan lab was to clone every known cytokine into the most efficient possible gene transfer vectors. Mulligan wasn't quite sure what to do with these constructs, but he at least wanted to start with the best possible tool kit.

Genetic tools in hand, Mulligan then suggested testing all the constructs on what are called B16 melanoma cells, "Which are the Mike Tyson of tumor cells," says Pardoll. They are totally nonimmunogenic—as far as the immune system is concerned, indestructible. If a genetic tweak with a cytokine had any effect on the immune system's response to B16 cells, that cytokine would be an excellent drug to take into further development.

After testing at least 20 different cytokines, Dranoff, Mulligan, Pardoll, and Elizabeth Jaffee (see Chapter 9), who was at the time a fellow in the Pardoll lab, came up with a winner: granulocyte-macrophage colony-stimulating factor (GM-CSF). "And right about the time when that work was being done, Ralph Steinman (see Chapter 10) published a landmark paper that demonstrated that the growth factor for dendritic cells [DCs, immune cells critical for vaccine function] was none other than GM-CSF," says Pardoll, still a bit amazed. "Everything kind of came together in one of those "Eureka!" moments."

GVAX

What came together was GVAX—not a drug exactly, but a therapeutic approach. The GVAX platform was a tumor cell genetically altered to produce GM-CSF. Theoretically, this approach should work with any tumor type. Working with a company founded by Dr. Mulligan called Somatix, Pardoll's first choice was to test GVAX against kidney cancer. It worked like this: (1) Take a biopsy from the patient; (2) transduce the patient's tumor cells with GM-CSF; and (3) vaccinate the patient with the transduced tumor cells. If all goes well, the GM-CSF signal alerts the immune system to the presence of the tumor.

There were two problems with this approach. First, the autologous nature of the protocol—using the patient's tumor rather than an established, off-the-shelf tumor cell line—was problematic and labor-intensive. Second, Somatix simply didn't have the cash flow to sustain the effort.

"But we'd taken it into the clinic and had some amazing anecdotal findings," says Pardoll, and enough attention was generated that a biotech company, Cell Genesys, acquired the vaccine. Cell Genesys, alas, proved to be no financial savior. For reasons unclear to Pardoll, the company quickly went with an alternate GVAX protocol for use in prostate cancer. "They made their decisions—there were some investor-driven things—you know how that works with biotech companies. And meanwhile, vaccines were failing right, left, and center."

There were two reasons for all the vaccine failures at the time, the first being straight-up ignorance. The basic science associated with DCs, the cells that inform and empower T cells to go after a target, just wasn't there.

The other reason that cancer vaccines failed is that tumors live in bad neighborhoods. "Tumors create a hostile microenvironment and that's where intratumoral checkpoints like PD-1 come into play," says Pardoll. It's an environment that's chock-full of signals and cells that suppress T-cell activity.

This begs the question as to why any vaccine works at all. The reason is this: The flu is not you. "Infectious disease vaccines are protecting individuals who have never seen the antigen," explains Pardoll. So, a flu bug is utterly foreign and easy to see. Further, infections are cleared by so-called neutralizing antibodies, which are produced by the B cells of the immune system. These antibodies are proteins that latch on to pathogens and render them harmless, neutralizing them. The mechanism whereby the immune system recognizes and destroys the much harder-to-detect cancer cell is different, involving both the provision of the target by dendritic cells and the resulting killing activity of T cells. However, little of that was known at the time.

Meanwhile, back to Cell Genesys, "A lot of people were putting a bet on this Phase III trial that Cell Genesys was doing with these prostate cancer vaccines."

It was a bad bet.

"I'm on my way to the airport to go to a meeting in Paris, and I take one last check of my e-mails and there's an e-mail from Cell Genesys's Chief Medical Officer. It said, 'Dear Drew, just to let you know that the trial was evaluated and the safety and monitoring committee decided that the probabilities of this being a positive trial were too low and we're stopping the trial and we're closing the company.'"

Enjoy your trip.

"Yeah, enjoy your trip," recites Pardoll. "So that was 2008, the absolutely lowest point of my career. I lost almost 20 pounds. I remember going to a meeting not long after that, a small meeting in Hilton Head and everybody—Steve Rosenberg, Carl June, Jim Allison—everybody like that was there, and I get up there and I say, 'Here's the question: Has our field finally died?'"

"Of course everybody else at the meeting was like, 'Come on now, we have successes, we've got this and that' and I'm like, 'Get your heads out of the sand.'" I said, "Steve, I don't deny that you've got some really interesting results, but it's never been exportable. I mean, we were at a meeting recently when somebody got up after you gave your spiel and said, 'Steve, this is the fifth time you've cured cancer.' I'm telling you nobody believes in us, nobody is investing in us, nobody cares."

"Nobody believes in us, nobody is investing in us, nobody cares."

What Went Wrong, and What Now?

Although still with hands shoved deep down in his pockets, kicking at the dirt, disgusted, Pardoll managed to reveal something shiny. It was a marvelous distraction and in the end, the beginning of the answer: Pardoll's lab had just discovered PD-L2, the sister molecule to PD-L1, a suppressive checkpoint of immune response (see Section II).

The discovery immediately prompted a question regarding the general scientific principles of discovery as posed years earlier by one of Pardoll's mentors, Don Coffey: "If this finding is true, what does it imply?" Pardoll was already thinking about the microenvironment of the tumor—all the little things you find there that you don't find in healthy, noninflamed tissues—and he asked the question: "If it is true that we were inducing T-cell responses by vaccines [various assays showed this], but that the response wasn't translating into clinical responses, what does it imply? Well, one of the implications is that the microenvironment of the tumor inhibits those T cells from doing anything."

This realization set off a cascade of related thoughts and activities: Pardoll's good friend and colleague, Lieping Chen, had discovered PD-L1 the previous year and then went on to show that PD-L1 was highly expressed in certain cancers; Japanese investigator Tasuku Honjo (see Chapter 4) had cloned PD-1, the receptor for PD-L1; Gordon Freeman (see Chapter 5) demonstrated the association between the PD-1 receptor and its ligand, PD-L1; and Honjo's knockout experiments in mice that resulted in auto-immune disorders showed the relationship of the PD-1 receptor and regulation of the immune system. "They got these [disorders] late and they were sort of tissue-specific," says Pardoll. "So all of a sudden we had a checkpoint that now was tumor-selective."

This was at the same time CTLA-4 was in the clinic, but (as discussed in previous chapters) the eventual approval of ipi was in great doubt. The Pfizer drug had crashed: Why would ipi be different?

The clinical program for the newer checkpoint inhibitor, anti-PD-1, was being spearheaded by Suzanne Topalian (see Chapter 6), who had conducted the first studies with ipi at the NIH. "Suzanne had started this first Phase I trial and reported the first few [positive] responses to anti-PD-1," says Pardoll, "Now, I had seen single responses to vaccines, so I was not impressed just because I was in a cynical, dark mood." But then came the encouraging results from the Phase II ipi trials, followed not long after by the extraordinary results from the Phase III studies.

How does this relate to vaccines? Well, the GVAX program is not dead. In fact, the GVAX program, as carried forward by Elizabeth Jaffee, is being tailored for cancer of the pancreas and is currently being tested in combination with the checkpoint inhibitor, nivolumab. Further, another protocol has GVAX being incubated with immune-stimulating cyclic dinucleotides (see Gajewski, Chapter 24) and is being tested by Aduro Biotech, in collaboration with the pharmaceutical giant, Novartis. "And guess who the Director of Oncology at Novartis is?" says Pardoll, grinning away, "None other than Glen Dranoff," one of GVAX's creators.

Small, Small World

If not already apparent in the reading of this book, the world of IO is small. Everybody pretty much knows everybody, and a not few of these bodies are married to each other, as are Drew Pardoll and Susan Topalian.

"We met over a slide projector," says Pardoll. Both he and Topalian had been invited to speak at the Keystone conference, organized that year by Mike Lotze (of the University of Pittsburgh) and held in the ski town of Taos, New Mexico. "Lotze always likes to say it was love at first slide." In this instance though, it was a slow burn. Admits Pardoll, "I'm not good at asking people out on a date."

As luck would have it (small world after all), a short time later Pardoll was invited to chair a think tank meeting on Hilton Head. "I accepted . . . only because it was an opportunity to invite Suzanne, and while I was on the phone with her I asked her out on a date."

Topalian said yes to the date, but the timing and the execution were a bit awkward. "It turned out that the last day she was at that Keystone meeting in Taos she fell skiing and fractured her pelvis." As for timing, "Suzanne was living in an apartment building right near the NIH, and I got her floor wrong," says Pardoll. "She was living on the 18th floor, and I went to the 17th floor and I knocked and there was this very sweet elderly Jewish couple that answered and they actually helped me find the apartment. So, I was an hour late. She was on crutches. That was our first date."

The rest is science history.

A Wrap with a Side of Allison

"We actually have an interesting history, Jim and I, including having dated the same woman," laughs Pardoll, adding for clarity, "but not at the same time." Yet, this light recollection hints at the pronounced competition in the field, a tension felt acutely when it comes to subject of funding.

"I'll tell you one of the most important stories about me and Jim Allison: The first grant I ever submitted was an R01 Grant for $212,000, and it got rejected," because it was an absurdly high amount for a first-time grant. "And I'm in the lab and the phone rings, so I pick it up and it's Jim. And he goes, 'You . . . idiot, Drew. I just got back from study session! Why the hell did you put this grant in for $212,000? It just pissed everybody off.' And he goes, 'This is what you're going to do. You're going to resubmit the same grant in the next cycle and you're going to put it in for $110,000.' I said, 'But I can't do it.' He goes, 'Shut up. You're just going to resubmit and you're going to put it in for $110,000.' And that's what I did and it got funded and it saved my ass."

Allison was on the funding committee and he and Pardoll were competitors in the field and Allison could have killed the grant without comment, "but literally in the airport on the way back from the study session he called me." There's a pause here, and a giggle. "Yeah, this field is full of great personalities." And all dedicated to one of the greatest of possible causes.

There are cures coming. Talk to any of these people and you will know. This dream is coming true.

ELIZABETH JAFFEE
"Mechanism of Action Hero"

CHAPTER NINE

Elizabeth Jaffee, M.D.

Deputy Director, Sidney Kimmel Comprehensive
Cancer Center
Co-Director, Cancer Immunology Program
Johns Hopkins University
Baltimore, Maryland

"Vaccines? Shouldn't you be doing something else?" —E. JAFFEE

Elizabeth Jaffee was born in 1959 in Brooklyn, New York, and for a while there, it was a blast.

"I loved living there!" says Jaffee, pushing forward in her chair, direct, speaking right into your face. "You walked everywhere. I walked to Hebrew school, to the library, to my favorite pizza shop. It was freedom!" The family home was modest at best, but it didn't matter. "Who knew you were poor? You didn't know you were poor." They had riches to spare; they owned the streets. "We played stoopball, yeah, that was big, and ring-a-levio."

Brooklyn Break: Ring-a-levio is a game like hide-and-seek, except with teams and jails. One team hides, the other team finds and imprisons. The playing field might be a block or the whole neighborhood. These games often were only called on account of darkness.

As a child, Jaffee wanted to be an astronaut, a dream eventually crushed by the harshness of certain realities. "It turns out I don't like tight spaces," Jaffee laughs, "And I don't like heights." Luckily, her fallback plan was to be a scientist, and that was a good thing because you should not dither about such things forever and she was already in the fourth grade.

"I read a very important book, the story of Marie Curie," says Jaffee, "And I just fell in love with the whole concept of doing science." Curie was a pioneer in the research of radioactivity, and the first woman to win the Nobel Prize. Units of radioactivity, curies, honor her name.

"I think I picked her because she was a woman scientist," but the choice had nothing to do with feminism. "In the fourth grade you don't really think that there [are] differences between men or women who go into science. I saw the challenge as doing science, not the challenge of being a woman in science."

That raises a curious question though: When does a little girl first become aware of sexism in science? Jaffee tries to find some humor in thinking about it. "I was my girls' Brownie troop leader when they were growing up, and one time I took them to a Science Day." It was an event featuring a group entirely of women scientists, ranging from a NASA scientist to a high school science teacher. They did these little experiments with the kids. "Then we sat around and each of the scientists told how they got into science, and then it was time for Q and A and one of the fourth-grade kids raises her hand and she said, 'Do boys go into science too?' And, 'if so, how do you work with them in the work- $\left(\quad \textit{"Do boys go into science too?"} \quad \right)$ place?' I mean, this is classic. And it just makes you realize that at that age girls think they can do anything."

So what happens? Why do so few of these fearless girls go into science? The answer to that question is likely a book in itself. "It's getting better, I think, but it's still an issue." And sexism is not just an issue for *little* girls. "When I started here [really not that long ago], they wanted me to start at $25,000 less than someone who was contributing less and starting at the same time. I started the whole translational program in immunotherapy. If I had left, I don't know that all this would have happened, at least not this quickly."

> *Note: At the time of this writing the Johns Hopkins webpage for Elizabeth Jaffee differs in one very interesting way from the page dedicated to her Hopkins' colleague Drew Pardoll (Chapter 8). Jaffee's page identifies her gender, as in: Professor of Oncology. Female.*

As her career moved forward, Jaffee speculates that part of her success as a women scientist can be attributed to her nature: demure, she is not. "I can be one of the guys. In fact, that was part of the problem growing up, . . . my interests were more [aligned] with men. I hung out more with guys in college; most of my friends were male. I had female friends, too, but I felt comfortable in that [male] environment. Maybe that's what it is," Jaffee suggests, pondering, looking about the room. "I felt comfortable and I didn't make people uncomfortable being a woman. It wasn't about woman or man. It was about 'Let's talk science.'"

It might also have had to do with her tastes, such as:

Star Trek *or* Star Wars?

"Now? I would say *Star Wars*, but I was a huge *Star Trek* fan."

Punk, New Wave, *or* Disco?

"Punk."

The Clash *or* Dead Kennedys?

"The Clash."

It's not a surprising answer. Jaffee favors short-cropped black leather jackets, and she wears her leather well. "Yes, I do."

Mentors and Beyond

Jaffee's first mentor was her fourth-grade teacher, Mrs. Freedman, followed in the 10th grade by Mrs. McDonald. "She was one of those nerdy old ladies with glasses. She was probably my age now, but she looked like an old lady," recalls Jaffee, "But she was just an amazing chemist. I thought, wow, here's this woman chemist, you know. She just inspired me."

After high school came college at Brandeis, where Jaffee's mentor was less inspiring. The time was 1977 and Jaffee had read a paper just the year before on hybridoma technology. It was a revelation and, with eyes opened maybe a bit too wide, she went looking for an immunology mentor.

"I wanted to use that technology to understand B cells, and I was working with this young faculty member at the time, Joan Press," says Jaffee, eyes narrowing. "She was one of these people who didn't like pre-med students, and she thought I just wanted to work there to get a good recommendation." Although mentor Press was willing to impart scientific knowledge, when it came to advising on career choices, "I was on my own."

So, her mentor wasn't mentoring, and there were no doctors in Jaffee's family to advise her. In fact, Jaffe was the first woman in the family to even go to college. Nevertheless, she got by. Perhaps it was the savvy of those Brooklyn streets. "I managed to navigate for myself and I ended up in medical school."

So, off to New York Medical College, then a year at the National Institutes of Health (NIH), then to Hopkins, and all during this time the concept of cancer immunotherapy was starting to play out as a real thing. "It was good timing because when I was in my residency we were just learning that IL-2 was good to grow T cells." (See Rosenberg, Chapter 13.) At the same time, other nascent tools and technologies were spurring the field, so much so that not long after she got into the show Jaffe was contributing to the parade of new techniques.

Innovations aside, initial support for her IO efforts was wobbly at best. "In fact, Mike Kastan, a good friend of mine . . . was meeting with me to discuss my career. He says, 'Okay, immunotherapy, yeah, you can do that,' but then he says, 'Vaccines? Shouldn't you be doing something else?'" The reputation for cancer vaccines at the time was that they were fundamentally flawed; tumors were simply thought to be nonimmunogenic. The immune system can't see them. Vaccines don't work.

"I probably should have listened," says Jaffee, "Who knows? Maybe I would have been rich and famous by now." But she didn't listen because the science made too much sense. "I really believed that vaccines were the best way to specifically activate T-cell and B-cell responses against any foreign antigen, and we were looking at cancer as making foreign antigens."

That was one of the first, and biggest, conceptual shifts bringing vaccines back into the light: the idea that genetic mutations in cancer cells could be seen as foreign. Yet, much of the rest of the related basic science was still waiting to be lured from the shadows. "What had to catch up were a couple of things," explains Jaffee. "One, we had to understand that there are a lot of checks and balances in the tumor microenvironment. [i.e., checkpoints]."

The second big shift in thinking, and one that is still very much underway, is that treating cancer according to tissue of origin—be it pancreas (Jaffee's specialty), lung, bladder, or whatever—is, at the molecular level, increasingly naïve. In fact, as the genetic data continue to pour in, not only is it obvious that tumors from one lung cancer patient differ greatly from those of another (and should not be treated the same way), but that the cells of a primary tumor (the seed from which the cancer grew) are often molecularly different from the distant metastatic tumors it spawns *in the same patient*. Further still, each patient's immune system is different in the way it reacts to cancer.

In aggregate, all this increasingly accessible information to characterize cancer types has implications regarding which antigens are to be used in a vaccine, how "personalized" a vaccine must be, and how a hostile tumor microenvironment might be therapeutically tamed prior to vaccination.

However, all of these ideas were still speculation at the time that Jaffee threw in with the "Vax-Heads." What prompted her choice were the new tools. "I got in as genetics was finally catching up," says Jaffee, explaining that many of the genes responsible for mediating the immune response had recently been sequenced, and genes, like ingredients in a cake recipe, can be used to bake all sorts of things. What Jaffee and colleagues Drew Pardoll, Hyam Levitsky, Glenn Dranoff, and Richard Mulligan baked was the first ever genetically engineered tumor vaccine.

GVAX

In a nutshell, the genetically engineered tumor vaccine, called GVAX, consists of live, irradiated tumor cells that have been transduced (i.e., genetically altered) with a cytokine: a protein-based signaling molecule called granulocyte-macrophage colony-stimulating-factor (GM-CSF) that acts as an immune system stimulator. Once the genetics are installed in vitro, the preparation is administered like any other vaccine.

The purpose of irradiating the cells is to prevent them from forming a new tumor after administration. The purpose of inserting the genetic instructions to make GM-CSF in tumor cells is to alert the dendritic cells of the immune system to take a long hard look at these tumor cells and figure out what's wrong with them, and then use the molecular signatures of those dysfunctions to instruct the killer T cells.

"You want to engage the patient's own dendritic cells," Jaffee explains, "Let them decide what to activate against. That's what GVAX is doing. GVAX is providing GM-CSF, which attracts and activates those dendritic cells. They decide what in the cancer cells is important and then they activate a T cell against it." This work took place in the mid-1990s.

"In the preclinical models [in mice] it was all working, and mechanistically it all made sense," says Jaffee, pushing glasses back, now looking through the frames. "So we had to figure out why it was when you went to patients, it wasn't working." Was the scientific connection between mice and humans lost in clinical translation? Many a compelling mouse data point has disappointed in that way, but in this case, Jaffee was not so sure. "There are certain things you can't get from mouse models, but there's a lot you can," and one of the things they got was proof that engaging the dendritic cell is important to activating a response against cancer.

Jaffe already knew from the work of Ralph Steinman—a good friend of hers who won the Nobel for isolating dendritic cells (Chapter 10)—that such cells were very important in recognizing the foreign antigens of infection, and she also knew many of the molecular mechanisms of how this happened. "It didn't make sense to me that I couldn't directly translate this from the mouse model against cancer." It was all there on paper, but it just wasn't working.

The why remained unknown for some time, but the where was a hot clue. "Lymphoid aggregates, yeah," says Jaffee, recalling the, *aha* moment. "So we gave GVAX two weeks before surgery and all of the sudden we're seeing these lymphoid aggregates coming in that look like lymph nodes in the cancer . . . Never saw that before." (Lymphoid aggregates are indications of an immune response. When your doctor feels your neck or in your armpits looking for swollen lymph nodes, she's looking for evidence of immune cells that have gathered in response to antigen recognition.) "It was clear the vaccine was doing something, and we saw this in 85% of the patients we treated."

Jaffee examined the aggregates and concluded that along with killer T cells there were cells and signals present in or near the tumor to keep the activity of the T cells in check. "They were being activated to see the cancer but they were being down-regulated, or they were being inhibited from moving out of those lymphoid aggregates because there were other signals. This was the definite aha moment."

The revelation was spatial, a clue to exactly where the suppressors of the immune response to cancer resided: in the tumor microenvironment. However, this information offered little guidance as to how to prevent immune suppression, and the related failure of the vaccine, from occurring.

Dark Night, and the Light

*Note: The demise of the GVAX initial clinical program, executed by
the biotech company Cell Genesys, is detailed in the previous chapter
highlighting Drew Pardoll, one of the co-developers of GVAX.*

"For me, it was bad," Jaffe recalls of that time. "I knew that it would be
hard to get grant money after that, but I had been funding the pancreatic cancer program myself anyway. The company wasn't funding for me because
they didn't believe pancreatic cancer would ever respond. [The company
was pursuing an indication in prostate cancer.] So when Cell Genesys fell
apart . . . yeah." It was tough, and it fueled the negative bias against vaccines
overall. "Everyone thought we failed," says Jaffee, and there was a tendency to
rub it in. "You know how the community is: 'Ha, ha, ha, it didn't work after all.
You thought you were so great. Blah, blah, blah, blah.'" It was 2008 and bio-
techs right and left were crashing.

"And Drew, too," says Jaffee. "He's a good buddy, and I know he can be a
little dramatic but yeah, it was a hit." For both investigators, the severity of the
hit was ameliorated by other projects. "It's not like we put all our eggs in one
basket. We both have big programs." It was just a matter of bucking up, dust-
ing off, and moving on. After all, there remained a near-endless stream of IO
questions to be answered.

And there were the patients. The
patients kept coming. "Patients love
immunotherapy. They've loved it from
the start. Even when the rest of the field didn't think we should be doing it,
patients wanted in on the studies," says Jaffee. It's pretty simple. Chemotherapy is harsh; everybody knows that. Vaccines, on the other hand, are not only
a very familiar technology, they also are relatively pain-free.

> "Patients love immunotherapy.
> They've loved it from the start."

For patients, it's an easily digestible conversation. "I explain to them that
what this vaccine is doing is it's activating their own immune system to see
their cancer and go on a seek-and-destroy mission throughout the body—anywhere in the body—so if there are metastases in the case of metastatic disease,
the immune system has the ability to seek and destroy that too." As to the procedure, the shots are more or less painless, side effects are minimal, and
immunotherapy helps you help yourself. It's an easy sell.

This is not to say that because patients are over the moon about immunotherapy, and investors are investing in the sector again (with caution),
that Jaffee somehow feels vindicated. That's because GVAX is no home run.

"I'll be vindicated," Jaffee starts off. Then starts again, "Vindicated is not
the right word . . . I will feel good when I have made an impact on pancreatic

cancer patients. She *will* feel: future tense. "I'm on the verge . . . ," Jaffee grapples with the timeline. "I particularly feel strongly now, more than ever . . . I've always been very passionate about what I do, but I couldn't do anything for my husband [Dr. Fred Brancati, who died from amyotrophic laterals sclerosis in 2013], but I feel like I'm on the verge of doing something for pancreatic cancer patients."

> *Note: Pancreatic cancer is one of the worst possible diagnoses; 71% of patients die within the first year. The average life expectancy after diagnosis with metastatic disease is around six months. Famous people who have died of cancer of the pancreas: Steve Jobs, Luciano Pavarotti, Sally Ride, Patrick Swayze.*

It's a heavy burden, and a weight she carries home. "I've known so many [patients] and the line can cross. They became friends." And not just the patients, there are bonds formed with family members. "It's painful what people go through, and I know what the pain is. I went through it with my husband. There was nobody there to help me, nobody in the medical field," says Jaffee. The finest medical professionals at Johns Hopkins said there was nothing to be done and that her husband would die, and that was that. "I don't want to do that to pancreatic cancer patients. I can give them hope and we are going to make a difference." Vindication is not the right word. "I will feel good. I will feel like I've contributed."

To be sure, Jaffee already feels pretty good about the notable exceptions to the usual outcome: There are some patients who have participated in GVAX studies who have done extremely well. For instance, Jaffee treated one patient for her invariably terminal cancer more than 15 years ago. "She comes one day and she says to me, 'You know, time for me to stop. I'm feeling good. But don't worry, I'll still call you guys.' So we brought a cake to the clinic and we celebrated."

Another patient, a man with advanced pancreatic cancer, who received GVAX as part of a clinical trial when he was 69, just died recently at the age of 91. "And his wife calls me and just thanks me," says Jaffee, "Like, he got all those years that she never thought he'd get. I want to see more of that, and really that's what gets me out of bed every day."

> *Note: The GVAX vaccine platform is now being investigated with the addition of a "prime boost" strategy at Aduro Biotech.*

FOUNDATIONAL DISCOVERIES, PROOF OF CONCEPT

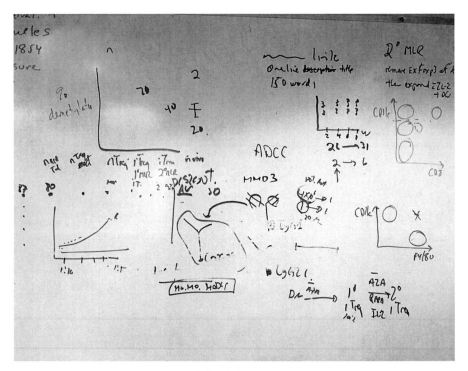

RALPH STEINMAN
"Ralph's Last Whiteboard"

CHAPTER TEN

Ralph Steinman, M.D.*

DENDRITIC CELLS

"No one believed him." — S. Schlesinger

After earning his medical degree from Harvard University, Ralph Steinman—born in Montreal, Quebec in 1943—spent his entire professional life in medical research at Rockefeller University, where he set up shop in 1972.

Meeting him there in the summer of 1976 was 16-year-old Sarah Schlesinger. She remembers: "He was tall. He was handsome. A little geeky . . . ," but very warm and kind. What brought Schlesinger to the Steinman lab were a lecture and a coincidence.

"The Rockefeller has a wonderful lecture series intended for high school students that's called, 'Talking Science'," says Schlesinger. Once called the Holiday Lectures because of their annual scheduling, the talks were inspired by, and modeled on, science demonstrations for children as performed in the early 1800s by the famous scientist, Michael Faraday (he of Faraday's constant, an electrical measurement). "The idea of the lectures was that if you get people in science when they're young, you get them for life."

That certainly was true for Schlesinger, now Dr. Sarah Schlesinger, Senior Attending Physician and Associate Professor of Clinical Investigation, Laboratory of Molecular Immunology, Rockefeller University.

The holiday lecture she heard was by Christian de Duve, winner of the Nobel Prize in Medicine in 1974, for his work on the microanatomy of the cell. "He did this wonderful lecture where we were cybernauts, and he took

---

*Dr. Steinman is deceased. His story is told by his long-time colleague, Sarah Schlesinger.

us on this tour of the inside of a living cell," recalls Schlesinger, still keen on the thought. "I can remember the lectures like they were yesterday."

"So anyway—this is Christmas time—I come home from the lecture and I say to my parents, I want to get a summer job there." A summer job at the age of 16, at Rockefeller University, one of the finest research institutions on the planet. "And they were like, okay . . . , but they had never heard of the place so it was unclear how I could make that happen."

As providence would have it, just days later Schlesinger's mother—a physical therapist who, with her family, lived north of the city in Westchester—attended a New Year's party with coworkers in Manhattan. During a flourish of cocktails and chitchat she encountered a woman, who had a friend, who had a husband, who was a doctor who worked ridiculously long hours even though he didn't even treat sick people (A doctor with no patients! Just imagine!) who worked at a place called Rockefeller University.

How very interesting, Mom thought, that a place she'd previously never heard of was prominently brought to her attention twice in the same week. The next day Mom told her daughter the interesting story. The daughter jumped on it. "I said, 'Mom, can you get me his phone number?'" Mom did. Schlesinger called the number. On the other end of the line was Ralph Steinman, the doctor with no patients (all his research was done in mice). The scientist and the teenager talked. The teenager got the job.

Ode to Claudia

Schlesinger arrived in the Steinman lab in the summer of 1977. The timing was perfect. She found herself on the ground floor of an argument being built to support a brand new discovery: dendritic cells. Steinman, with support from his mentor, Zanvil Cohn (he of the Zanvil Cohn Vaccine Center) was trying to convince the scientific world that his discovery, some new thing called a dendritic cell, first described in a paper published in 1973, was real and it was important.

"No one believed him except Dr. Cohn," says Schlesinger. "Everybody said they were macrophages." This was not an unreasonable pushback. A macrophage is a type of immune cell that gobbles up foreign things. It's kind of blobby looking, and it undulates and reaches out to its surroundings with pseudopodia—little appendages that shorten or lengthen at will—and it communicates with the wider immune system. The proposed new cell, the dendritic cell, did all of that and looked pretty much like that, and further, they were isolated in the same way.

It also didn't help that Steinman's initial pursuit—the reason he came across dendritic cells in the first place—was that he was investigating the function of macrophages. "Ralph came to Rockefeller to learn how immunity was initiated," says Schlesinger, "That was the question."

It was already known that B cells and macrophages could present antigen (bits of the foreign thing that had been gobbled up by the cells) and they could elicit a recall response by the immune system, but how that initial *primary* response was initiated was not understood. Steinman was trying to understand it.

To that end, as with most biological investigations, one must first gather up a great pile of the thing to be investigated. In this case, the gathering and isolation of macrophages went like this: Get a mouse. Wring its neck. Yank out its spleen and mash it up. After mashing, spike the spleen preparation with ammonium chloride to lyse the unwanted red blood cells present in the mix, and then culture the whole mess overnight in the presence of serum in a glass dish. The next morning, all the macrophages are stuck to the glass.

Simple.

"Then you blow off the other [hangers-on] cells with a pipette and you're left with the adherent cells," Schlesinger explains. The adherent cells are cultured again overnight, and those that remain stuck to the glass are a pure sample of healthy macrophages.

For Steinman, however, there was a next step—the discovery step—and that involved a great deal of tunnel vision, the "tunnel" in this case being the barrel of a microscope. "If you looked carefully through the microscope the cells that had let go [after culturing] were different from the cells that were adherent," says Schlesinger.

These preparations were not unique to the Steinman lab, but apparently —for as long as researchers had been making them—no one had bothered to take a close look at anything but the cells that remained stuck.

Microscopes focus at a certain depth, be it 1 micron, or 10, or 100, and the only things you see are what exist at that focal depth (1 inch = 25,400 microns). "Because we're talking about focusing up and down with a phase microscope, Ralph looked at the ones that were floating and noticed that they were morphologically different." There was a distinctive look, similar to macrophages, but not identical. "Those are the dendritic cells."

As stated in that first paper by Steinman and Cohn in 1973: "We noticed a large stellate cell with distinct properties." The paper goes on to reveal a bevy of micrographic images of this new cell type, its pseudopodia elongating this way and that, as well as details

> "We noticed a large stellate cell with distinct properties."
> —R. STEINMAN AND Z. COHN

regarding where the cells were found: intestine, thymus, bone marrow, lymph node, liver—all places where immune cells typically like to lounge. The paper also proposed a name for the cells with the branched, limb-like projections. Taking from the Greek *Dendron*, meaning "tree," Steinman coined the term, dendritic cells.

Although that was not his first choice.

"Ralph wanted to name them after his wife, Claudia," says Schlesinger. "She's a lovely woman. She's very slender and has long arms and legs [kind of like dendrites] and so he wanted to name them Claudiacytes." For her part, Claudia Steinman was flattered by the nerdishly romantic gesture but convinced her husband that such an honor was not precisely in the interest of science. "So, 35 years after I first attended, I got to give the Holiday Lecture in Ralph's memory, and Claudia was there. She told me this wonderful story about accompanying him to meetings and people looking at her arms and legs because he talked about it so much." It was obvious.

"He adored her."

Skin Game

With the cells having a safe, classically academic name and a documented distinctive shape, the next step was to determine what they did. The answer was slowly and methodically produced by Steinman and his, by then, growing team. Schlesinger worked multiple summers in the Steinman lab between bouts of formal education, eventually earning her medical degree at Rush Medical College.

First, a general association was established: Dendritic cells bind directly to one of the most effective killers in the immune system—the T cells. This capacity was revealed through a very simple assay called the mixed lymphocyte reaction (MLR), whereby T cells are activated and proliferate greatly in the presence of dendritic cells. At the time these data were produced, the assay required T cells isolated with fresh sheep's blood.

"You'd take sheep red blood cells and mix them with leukocytes," explains Schlesinger. Then spin down the concoction on a centrifuge, add a splash of ammonium chloride—just like the spleen prep—and what you're left with are T cells.

Note: The MLR works because sheep red blood cells express a molecule called CD1, a molecule also expressed on antigen-presenting cells such as dendritic cells. T cells bind naturally to CD1.

"So yeah, there was a sheep here [at Rockefeller]. I can't remember the name of the sheep just now, but that was part of my job: to go get the sheep blood."

To review: the skills learned by Sarah Schlesinger, intrepid summer intern, were sheep bleeding, mouse wringing, blending spleen smoothies, and babysitting swatches of human skin.

To explain: When Schlesinger first came to the lab, the standard method of attaining *human* dendritic cells was to coax them from one of their natural organic reservoirs, the skin. Your skin is the secure border between you and a world full of pathogens. It makes sense then that dendritic cells, as was later fully realized, are the guards lined up just behind that border. This also explains how vaccinations—that squirt of foreign matter just under the skin, which is then taken up by dendritic cells—accomplish their goal of protection against disease.

"I got human cadaveric skin, like the kind that people get for skin grafts [burn victims, for example] and the dendritic cells grew out . . . they crawled out," says Schlesinger. "It was extraordinarily taxing to do. You're spending three days to get 10,000 cells, if you're lucky. It's very hard, very expensive, and skin is fundamentally a contaminated tissue," so, all around, fiddling with skin is a pain in the ass.

Which is why so few people did it. The skin thing, the sheep, the spleen preps, the tedious hours of microscopic observation: The methods were too crude, and the results too arbitrary. For example, if someone did manage to isolate what they thought were dendritic cells, how could they prove it? "In the beginning, the way you knew it was a dendritic cell was that Ralph said so. Truly. I mean, it was strictly morphologic. We'd look at them and you'd have to get Ralph to say, 'Yes.'" There was no biomarker, no objective method of identification. Just Ralph.

"Ralph loved to look in the microscope," says Schlesinger. "Right up until his death he looked in the microscope at those cells. Some of my happiest times with him were across a two-headed microscope just enjoying the pleasure of looking at those cells. They are quite beautiful."

Meanwhile, Steinman's team demonstrated a key element of dendritic cell function: They could display antigen. This is exactly how T cells knew what to attack. A dendritic cell engulfs a foreign thing, chops it to bits, and then decorates itself with the best bits. T cells see the bits and then go after anything that resembles them.

Even with that progress, Ralph was still the singular authority on dendritic cells. He was the one to give thumbs-up or -down on your findings. That rankled, and it most certainly motivated some of the pushback against this new proposed cell type.

However, in the late 1970s and early 1980s, two fundamental developments in science leveled the barrier of entry to the field, and after that, most anyone got to play. Those two developments were **hybridomas** and recombinant cytokines.

Hybridomas: A technology that uses living cells to produce unlimited amounts of highly specific antibodies, all exactly the same (i.e., clones). One of many uses for such cloned antibodies are the identification and purification of a given target molecule or cell type, such as T cells (thus eliminating the need for sheep's blood or spleen smoothies).

Recombinant cytokines: Recombinant proteins (cytokines are proteins) are so named because scientists recombine existing strands of DNA with other DNA to facilitate the large-scale manufacture of a given protein in a bioreactor. This is common practice in the pharmaceutical industry. (Medical insulin, a protein-based hormone for diabetes patients, is made this way. Before the advent of recombinant technology, insulin for human use was extracted from the pancreas of pigs.)

Recombinant cytokines proved to be the way to make mature dendritic cells from their parent progenitor cells (i.e., cells found in the bloodstream and bone marrow that, when properly instructed by cytokines, mature into a virtually unlimited supply of dendritic cells).

Despite these advances, controversy remained in the field because there was still no standardized way to determine if the cells coaxed into being using recombinant cytokines were the same as dendritic cells derived naturally (free-range, if you will).

These uncertainties and disagreements were finally put to rest in 1998 when Steinman and his coauthor, Jacques Banchereau (of Baylor) published a paper in *Nature* that spelled it all out. As the first sentence of the abstract states conclusively: "B and T lymphocytes are the mediators of immunity, but their function is under the control of dendritic cells." The next two sentences stated exactly how dendritic cells accomplish the fact stated in the first sentence, while the fourth and fifth sentences offered this: "Once a neglected cell type, dendritic cells can now be readily obtained in sufficient quantities to allow molecular and cell biological analysis. With knowledge comes the realization that these cells are a powerful tool for manipulating the immune system."

The paper goes on to include descriptions of all the materials and methods needed for any scientist to join in the ongoing investigations.

Suddenly—25 years later—the study of dendritic cells became a bona fide wide-open scientific field rather than Steinman's private garden. Thereafter came the push toward clinical translation of this seminal discovery.

Dendritic Cells in Prime Time

Between the summer internship in 1977 and the *Nature* paper of 1998, Sarah Schlesinger kept busy. Her activities in brief:

- Medical degree at Rush University: "I started school in 1981, the year of the first reported AIDS case."

- A stint in surgery training, followed by realization that the rigors of a surgeon's life were not for her. "The notion that there weren't enough hours in the day had never struck me before."

- Two marriages.

- Five kids, all boys, the last in 2011. "It's a long story."

- A specialty in pathology. "Ralph, in his unending generosity, helped me get a position in pathology at New York Hospital, and pathology is a tried and true specialty for people who want to be in the lab."

- An eight-year stint at Walter Reed Army Institute of Research working on vaccines for HIV.

"People at Walter Reed were starting to become interested in dendritic cells," says Schlesinger, "And I had learned at the hands of the master." Yet, there were things the master did not know, and what Schlesinger learned at Walter Reed would eventually play a significant role in the Steinman lab.

"I was required to learn certain things because I was a medical officer—good laboratory practices, manufacturing, clinical development, regulatory affairs—all these things that a medical officer in the Army has to know." Those mandated skills prepared her for a life of translating scientific discoveries, such as those of Steinman, into clinical reality.

"Ralph, who had never really been interested in doing anything in humans because he was a mouse doctor, . . . realized that people were starting to use dendritic cells clinically," recalls Schlesinger. Steinman was not entirely crazy about the ways colleagues were dandling his baby, manipulating the new cell type haphazardly (in his opinion) for a myriad of therapeutic applications.

But the genie was out of the bottle. At first, there was a great deal of enthusiasm (and money) for this leveraging of immune cells for cancer immunotherapy (all through the 1990s, in fact) but the projected clinical fireworks all but fizzled. "Ralph was extremely critical of the people involved in those experiments, including some who purported to use dendritic cells," says Schlesinger. As far as Steinman was concerned, the scientists had gotten ahead of the science.

"He always thought he could do things better than everybody else. He wasn't usually wrong." He wanted to control, as best he could, the clinical

"He always thought he could do things better than everybody else. He wasn't usually wrong."

application of dendritic cells. Steinman wanted to begin treating patients. "So I came back to be clinical director of his laboratory."

That was 2002.

Schlesinger's return was not merely a testimony in loving support of the lab or its leader, although there was plenty of that. Rather, Schlesinger the scientist had been duly impressed by several publications from the lab in the late 1990s, particularly those executed by lab members Nina Bhardwaj (now at Mt. Sinai) and Madhav Dhodapkar (now at Yale).

"They started doing very, very basic experiments using dendritic cells to demonstrate that you could immunize somebody with flu, tetanus, and KLH [a marine antigen]," says Schlesinger. The experiments were sheer elegance in their simplicity: Perform leukapheresis on a patient (i.e., filter out their white blood cells), subject those cells to cytokines, grow out the dendritic cells, then load the dendritic cells with peptide antigen in vitro and return the now primed and activated cells to the patient.

It was loading dendritic cells with antigen that was the neat trick. Steinman's group was recreating in a glass dish what normally happens in the skin, or a lymph node, or, indeed, anywhere in the body where the immune system might encounter a pathogen.

With this simple method, Steinman was able to elicit an immune response in patients as if he had given a vaccine. "That was the first indication that I had that dendritic cells really could be manipulated," said Schlesinger. "Those papers were, frankly, the things that said to me: 'Wow!'"

These data were impressive enough that Schlesinger applied those methods, and her already extensive dendritic cell know-how, to the ongoing HIV vaccine program at Walter Reed with which she was involved. That work was developed further after her return to Rockefeller. There, under Schlesinger's guidance, clinical testing of an HIV vaccine platform continues to this day.

But that's now. Back in 2002, Schlesinger returned to the Steinman lab and for five years enjoyed a steady stream of funding, and related scientific progress.

In 2007, Ralph Steinman got sick. He was diagnosed with pancreatic cancer, with a grade 3–sized mass at the head of his pancreas. The one-year survival rate for such a diagnosis is less than 4%.

"So," says Schlesinger, "That changed everything."

Steinman became the experiment.

The Grand Experiment

Calls came in from all over the world. Close friends, colleagues, leading scientists each and all, reached out.

"Ralph was dearly loved, and he had a huge outpouring of support and caring and offers. Basically, anybody who had anything that might be of use offered to provide it," and we're not talking about warm wishes and casseroles here; the technology being volunteered was fresh out of the world-class labs where it was just invented.

Schlesinger became the default coordinator of the effort, a role familiar to her from her years with Steinman. "Ralph [convened] these meetings that were sort of like lab meetings with maybe six people, and some on the phone." They brainstormed. Ideas were put forth, priorities determined, timing and logistics discussed.

Schlesinger took the notes. A treatment plan was devised. There was only one thing left to decide. "After everybody left I was still organizing papers and I said to Ralph, 'So, who's going to treat you? Who is actually going to administer these therapies?' And that's when he said to me, 'Well, I would like you to.'"

It was a profoundly touching request, but not exclusively so. "My first reaction was surprise, because I'm not an oncologist," says Schlesinger. She instead recommended several colleagues who were perhaps better-suited to the task. But Steinman was gently insistent, and in retrospect, Schlesinger thinks she knows why. "I was always scientifically deferential to Ralph, and I think he felt that I, perhaps more than others, would be willing to follow his guidance and judgment in preference to my own."

Because, bottom line, regardless of the identity of the test subject, Steinman insisted on running the experiment.

"Absolutely. He was in charge, and our relationship was always me as his deputy; that was our relationship." That said, Steinman wasn't stupid. Schlesinger was the unchallenged clinical expert of the two, and her hard-won acumen was indispensable for surmounting the regulatory and institutional obstacles unique to the project at hand.

After the plan was set and the path cleared, treatment commenced. First up: GVAX, a cancer vaccine technology pioneered by Elizabeth Jaffee (of Johns Hopkins; Chapter 9) and Glen Dranoff (then at Dana–Farber, now at Novartis).

"Liz made a special GVAX which was Ralph's cells [tweaked] with the exact same process that she used for the general GVAX, and we got permission from the FDA to use that." The vaccine was administered in the fall of 2007.

Steinman had a measurable response to treatment; biomarkers detected in his blood indicated that T cells had been activated. What effect the activated T cells had on residual cancer cells Ralph had after his initial surgery were unknown and could not be known given contemporary methodologies.

Yet, Steinman forged ahead. Next up in the winter of 2007 was a dendritic cell–based vaccine prepared by colleagues at a company cofounded by Steinman called Argos. In the spring of 2008, Steinman was treated with a dendritic cell vaccine prepared for him at Baylor University by Karolina Palucka, with input from Jacques Banchereau, Steinman's coauthor from the seminal *Nature* paper.

"We were administering the [Baylor] vaccine to him when Karolina came up here to New York," Schlesinger recalls. Steinman had received his vaccine the previous day, injected in the front aspect of his thigh. The following evening, the three of them went out to dinner. "And he was so excited because he was getting a very robust, local immune response," describing the injection site to his dinner companions as red, and indurate (hard), "which meant that there were T cells there."

"He insisted that we go to Karolina's hotel room so that he could show us, and Karolina . . . she knew Ralph very well but didn't quite know what to think of it." It was Ralph Steinman after all, a friend, a colleague, a mentor, but it was High Science too. "I have to say, I too was a little overwhelmed, but we all sort of got into it and were excited. If there had been smart phones back then, we probably would have taken pictures of it."

There was another medical intervention or two after that hotel visit (including the use of ipilimumab; Chapter 1) but four-and-a-half years after it started, on September 30, 2011, the Grand Experiment came to an end.

✤ ✤ ✤

September 30, 2011, was a Friday.

That Friday, Steinman's daughter called Schlesinger with the bad news. Her father had passed. The daughter then asked Schlesinger to refrain from telling anyone else until the coming Monday, to give the family time to catch their breath before the onslaught of the well-meaning.

"I understood completely."

Saturday came and went, as did Sunday. "And I went to bed Sunday night, but I didn't sleep well because I knew that I was going to have to go into work and break the news to everybody," so Schlesinger was more or less conscious at 5 a.m. Monday when her phone started madly yipping. Schlesinger looked at the screen. It was Steinman's daughter.

"I picked up the phone and she said, 'Dad won'. And I said, 'Honey, didn't your . . . ,' I was like half asleep. I said, 'Didn't your dad die? Am I remembering

> *"I picked up the phone and she said, 'Dad won.'"*

right?' And she said 'No, no, no, of course. Yes, he died . . . he died two days ago . . . but the Nobel people are calling to tell us he won the Nobel Prize!'"

Vitally Important Note: The rules of the Nobel Prize state that the award cannot be given posthumously.

Steinman's daughter was in a panic. What do I do? Who do I call? "I said, I have no idea, but I'll find out."

Long story short: After consulting with lawyers and alums and decent people all around, the Nobel committee decided to make an exception to the rule and released the following statement: "The Nobel Prize to Ralph Steinman was made in good faith, based on the assumption that the Nobel laureate was alive."

Weeks later, at the official awards ceremony in Stockholm, the 2011 Nobel Prize in Physiology or Medicine was awarded to the late Ralph Steinman, and accepted by his cherished wife, Claudia.

"The only way to describe it is 'surreal'," says Schlesinger. "We used the word 'bittersweet' so many times that it was really a cliché in that period of our life, all of us." Steinman had reached the Everest of scientific success and everyone in the world knew it, except for him.

"We accompanied his family to the ceremony, and yes, I watched with tears in my eyes as Claudia accepted the Prize on his behalf. And then after that I got to hold it, and I looked at it, and caressed it more times than I like to admit. It made me so sad that Ralph missed it. He would have been so happy!"

✤ ✤ ✤

Madhav Dhodapkar, M.B.B.S.

(Former trainee, Steinman lab)
Professor of Immunobiology
Yale University

"His passion came out instantaneously. He even did science when he was very sick. You would get e-mails from him when he was just barely out of his surgery and he was talking about real scientific questions. It was almost a childlike curiosity . . . That just blew me away."

✤ ✤ ✤

Drew Pardoll, M.D., Ph.D. (Chapter 8)

"Yeah, they don't give the Nobel Prize posthumously, so there was this whole thing, like, are they going to actually retract it because he's dead? I can tell you it would have caused a shitshow if they did, because he was really quite revered by the community."

❀ ❀ ❀

Jacqueline Chiappetta

Executive Secretary to Ralph Steinman
New Yorker, Born and Raised

Note: Chiappetta has recently retired. One of the many guests she ushered into her boss's office over the years was a 16-year-old Sarah Schlesinger.

"He always came in early at the crack of dawn, and maybe he left the office, I would say, 7 o'clock and then he worked at home. He worked so hard."

"He was a good boss, but very demanding. I used to find in the morning when I woke up that I had little yellow stickums all over my slippers, because I would think of things I had to tell him in the morning because he was so busy. And I'd get up and I'd have all these yellow stickums on the floor. But he so appreciated everything we did for him."

What was your reaction when you found out Dr. Steinman was sick?

"We all tried to be normal, but I did one thing I never did before in my life. I started getting the flu shot because I didn't want to get sick and expose him to more germs. I never did it before. He got sick and I started getting the flu shot."

And after he passed?

"It was really hard coming to work because my office was right outside of Ralph's. His wife couldn't deal with like, cleaning the office or anything at first, so even his shoes were still on the floor and his dirty cup of coffee was on the desk."

"That's really sad that he didn't get to see himself win the Nobel Prize. I don't think it would have changed his demeanor at all, but I would have liked him to get that little pat on the back."

❀ ❀ ❀

Elizabeth Jaffee, M.D. (Chapter 9)

"His wife and his lab had a wonderful memorial about two weeks later; it was a celebration of his life, which I thought was just so beautiful. Everybody who had worked with him, those he mentored, I mean, they all came."

"He was really into salsa, him and his wife, so the place was packed and there was a salsa band. It was just everyone saying things about him, nice things about him . . . Then it hit me. You know what? It didn't matter that he didn't know he won the Nobel because look at this: just look. This is really his legacy."

Endnote: To this day, there is no effective treatment for late-stage pancreatic cancer. Steinman was treated with six different experimental protocols for a disease that should have killed him within the first year. Instead, he lived four and a half years, but that doesn't mean the treatments worked. No one involved in Dr. Steinman's care can scientifically say if any of the cutting-edge interventions he underwent actually did any good. Sometimes terminally ill patients live a long time for no particular reason.

Nevertheless, one researcher was convinced. One researcher knew that there had been a benefit, and that they were on to something right, and true, and trailblazing. That researcher was Ralph Marvin Steinman, M.D.

Mouse Doctor.

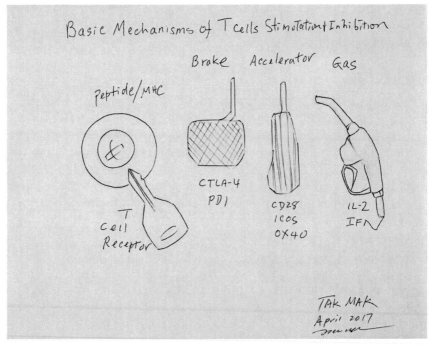

TAK MAK
"Gentlemen, Start Your Engines"

CHAPTER ELEVEN

Tak Mak, Ph.D.

Senior Scientist
Princess Margaret Cancer Centre
Toronto, Ontario, Canada

T-CELL RECEPTOR (TCR)

"Unless I made it up, it's right." —T. MAK

The essential knowledge of Tak Mak includes three things.

- Tak Mak discovered the T-cell receptor (TCR), a seminal discovery that serves as the molecular foundation for any number of technologies described in this book. The importance of his achievement cannot be overstated.

- Tak Mak is funny, the kind of funny that you need to be in the same room with to fully notice and appreciate: master as he is of the subtle eye roll; the timely, if only slightly raised eyebrow; and the whispered aside contradicting what, in full voice, he just said. He sets the listener immediately at ease with this demeanor and perhaps this is deliberate, as suggested by his close friend, James Allison (see Chapter 1): "Tak hides his scientific brilliance in humor."

- Tak Mak speaks just as softly as he smiles. He is obviously open, warm, and generous. In a word, Tak Mak is Canadian.

✤ ✤ ✤

Tak Mak was born in 1946 on the island of Hong Kong. His early education, until he left for college in the United States after high school, was at the hands of Irish Jesuits. "It's perhaps the most scholarly brotherhood of the different Catholic denominations," says Mak. "They believe in knowledge and, of course, they believe in God, and everyone was always very kind and gentle."

In addition to imparting their love of learning and their bent toward compassion, the Jesuits also tutored Mak on the burdens of the faith or at least the cultural expression thereof. "There's an old saying: 'The Chinese and the Jews are *born* with guilt, but the Catholics learn it at school.' And guess what?" Mak smiles. "I am a Chinese Catholic." Thus, Mak was born with guilt, and the Jesuits fed and groomed it throughout his youth. It shaped him. "It's like, everything I do now, in the back of my mind, I do maybe . . . maybe because a guilt feeling can be translated into other kind of feelings, like empathy, like sincerity, like charity."

This enlightened take on a heavy weight plays out in any number of ways in Mak's life, not the least of which are his ongoing contributions to the Croucher Foundation, a private organization founded in 1980 to, according to their website, "promote the standard of the natural sciences, technology, and medicine in Hong Kong." Among other supportive activities, like hosting conferences and workshops, Croucher provides scholarships for fledgling Hong Kong scientists—nearly a thousand such awards to date.

The first president of the Croucher Foundation was the British-born, University of Oxford graduate and Nobel Prize winner, Lord Alexander Todd. The current chair of the Board of Trustees of the Croucher Foundation, since 2011, is Tak Mak.

Although his long-time home is Canada, Mak maintains a strong affinity for the people of Hong Kong. "They are very different from people in China," says Mak, as different as they are from residents of the other blockbuster island, Singapore. To wit: "People from Singapore are Chinese who wish they were Swiss, whereas the people from Hong Kong are Chinese who wish they were New Yorkers."

The upside of this is that everyone in the society is pretty much on the same cultural page. "It's a practical kind of integrity. Probably the city with the lowest crime rate in the world is Hong Kong. There is no such thing as crime, and that's because people all work together to build the future," says Mak. He adds that there is a bit of a downside to this homogeneous identification in that, just like New Yorkers, the focus on building the future translates into building piles of money.

So be it. More funding for science.

Blessed Are the Cheeseheads

After high school, the Mak family—the boy, two sisters, and a mom—moved to Madison, Wisconsin, so that the children could go to the university there, and so that Mak could study to be a doctor. At least, that's what his mother thought. "Jewish parents, Chinese parents—doctors/lawyers, same thing. There's so many parallels between Chinese and Jews," says Mak. So as not to disappoint right away, he didn't tell his mom until the wheels were already in motion that he would be majoring in chemical engineering.

That decision didn't last. A year in, Mak decided to ditch chemical engineering, the stuff of plastics, in favor of biochemistry, the stuff of life. It was a good fit, and all according to plan. (Insert Mak-whispered aside here.)

"I needed to pay for my extra expenses at school so I applied for a job from the Student Union," says Mak, "I started as a construction worker, but at that time I was 110 pounds so there was no way I could really do that. After a few tries, I decided, well, I'm just going to have to wash dishes." But not just any dishes: Mak wanted to wash laboratory dishes. He targeted two labs, those of the botanist Folke Skoog and of Roland Rueckert, a virologist. "I was curious about what they were doing, and in both cases they tried to entice me to [join in]." Dr. Rueckert won out, offering to pay the munificent sum of $1.25 an hour (minimum wage in the United States at the time, 1965). "The first time I washed dishes I finished in two hours, but that wasn't enough money, so I asked, 'Are there any more dishes?' and Rueckert said 'No, but if you do experiments I'll pay you $1.50 an hour!'"

"So he made me this offer I couldn't refuse, and then I ended up basically doing experiments with him . . . first for him, and then with him." Before long, Mak changed his major to biochemistry and signed up to do his graduate training in Rueckert's lab.

> "So he made me this offer I couldn't refuse."

Mak earned his B.S. in biochemistry in 1967 at the University of Wisconsin. Following that, he switched gears and earned a master's degree in biophysics at the same institution.

After the master's came a Ph.D. in biochemistry in 1972 at the University of Alberta, followed by a postdoc in Toronto at the Ontario Cancer Institute, a division of the Princess Margaret Cancer Centre, where Mak worked under the guidance of preeminent scientist, Ernest McCulloch.

It was McCulloch who would safeguard Mak as he searched for his place in science. It was McCulloch who, along with his co-investigator James Till—just before Mak's arrival in the laboratory—discovered the massively clinically important thing called a hematopoietic stem cell, the progenitor cell that resides in the bone marrow, the wellspring from which all the cells in the bloodstream are derived: T cells, B cells, red blood cells, everything.

Till and McCulloch should have won a Nobel Prize for their work.

They did not.

"So, just to give you an idea of how this guilt thing works," explains Mak, "The team of McCulloch and Till missed the Nobel Prize because a scientist named Sir John Gurdon [did an experiment] transferring the nucleus from a fibroblast cell into an ovary cell, and then demonstrated that that cell, a stem cell, can develop into a new mouse." Essentially, Gurdon made a baby from an adult. It was a hell of a trick, and Gurdon didn't even need to know all the details of the trick to perform it, whereas McCulloch and Till, who were fiddling with much of the same stuff wanted to nail down all the molecular details before making their work widely known.

It was error on the side of scientific caution, but a career fumble none-theless.

"They were waiting for a mechanism to explain it," says Mak, and in the meantime, Sir Gurdon snagged the Prize. "But make no mistake, McCulloch and Till discovered hematopoietic stem cells right here in this institute in 1961." (There is near consensus on this point in the scientific community, but the opinion is not universal.) "Anyway, McCulloch passed away about five years ago, and I wrote to the Royal Society's *Philosophical Transactions* [the world's first science journal, published since 1662] and I said I want to write a memoir of McCulloch because . . . Well, all the time it bothered me." It bothered him that his mentor, McCulloch (whose picture is still on Mak's desk to this very day) did not receive the recognition that Mak thought he deserved. "I feel that I owe him because he was the guy who gave me the chance," says Mak—the chance and the vital support to pursue what was known in the field at that time as the "Holy Grail of Immunology": the cloning of the TCR.

Corralling the Crazy: The Discovery of the T-Cell Receptor

What Mak was up against—and why he needed McCulloch as a mentor and defender—was his idea that, at the receptor level, T cells and B cells are completely separate entities. This assertion was anathema to immunologists. Much was already known about B cells, so scientists were feverishly (and logi-cally, they thought) looking for the TCR based on what they knew of the design of the receptor on B cells. They figured there had to be a physical relationship, a similarity between the two, "because nobody thought there was enough time for nature to have evolved a completely different system of antigen recognition."

Young Mak thought otherwise. He thought the TCR would be unique, and further, because of the techniques he learned in his virology experiments, he thought he knew how to find it.

The technique most valuable to the search is called subtractive hybridization, a method whereby you compare two sets of similar but distinct genetic instructions (like those of B cells and T cells) and you subtract one from the other to see what's left, (i.e., all the genetic distinctions between the two).

The proof of principle, the example of subtractive hybridization from which Mak was operating was the discovery in 1970 of the first oncogene, a genetic mutation that causes cancer. That earlier work (done by Varmus, Stehelin, Bishop, and Vogt) involved the comparison of two types of virus: the avian leukosis virus, and the Rous sarcoma virus. Under certain conditions, both viruses caused cancer in chickens. What Varmus et al. wanted to know was the causative agent: What was the gene that, when inserted into a healthy cell (which is what viruses do), turned it cancerous?

The experiment was actually quite simple. "In the Rous sarcoma virus, you have just four genes," explains Mak. "They are called *gag*, *pol*, *env*, and *src* [the last pronounced, "sarc" for sarcoma]." The genes encode proteins for everything the virus needs to reproduce. The leukosis virus, on the other hand, has only three genes: *gag*, *pol*, and *env*. "They used subtraction hybridization to subtract away [the genetic differences] and they ended up with the gene for *src*, and they got the Nobel Prize for it because they showed that, after all, *src* was a normal chicken gene, but it was just mutated." That's where the cancer came from, the *src* oncogene.

"So, I went to my boss, McCulloch, and I said, look, if it's taking so long for people to clone the T-cell receptor [years, in fact, making it the Holy Grail] maybe it is not part of B cells. And if it's not part of B cells, then I could do subtractive hybridization." What could be simpler? For the *src* experiment, it was four minus three. To compare the genes of B and T cells it would merely be, say, 7000 minus 6800.

(Slight raise of the eyebrow.)

"That was a crazy idea," says Mak, but mentor McCulloch had his back and told him to write up and submit a grant, which Mak did; the granting committee promptly turned it down. Reasons given for the rejection were largely issues of audacity and geography.

For audacity, one need consider the heft of the competition: Lee Hood, the genetic sequencing giant out of Caltech, had 56 people in the lab working on the problem. "They pretty much had the whole building to themselves," says Mak. At the same time, Nobel laureate David Baltimore, also at Caltech, had at least 20 postdocs, "and Mark Davis, he was in Bill Paul's lab, and Bill Paul had an empire at the NIH."

And then there was Tak Mak, a lone postdoc at a small lab in Canada.

"I was two years into a postdoc—and Canadian—and they basically said to me, 'What the hell are you doing? This is a rat race; it's not for you.'" The prevail-

> *'They basically said to me, 'What the hell are you doing? This is a rat race; it's not for you.'"*

ing sense was that the project was too big, too bold, and frankly, no Canadian should be involved in such a venture, an attitude Mak still wonders about. "We don't have a history of big dreams, I guess."

However, to be fair to the granting committee, the technical aspects of the project were beyond daunting.

"Even though you can do four genes minus three," says Mak, "Nobody had ever attempted doing 7000 minus 6800." Moreover, given the proposed method, the goal seemed impossible. What Mak wanted to do would take days, as opposed to the two hours it took to do the four minus three experiment, and because the material being used in the assay was RNA, which is inherently unstable, the educated assumption was that the RNA would degrade long before the hybridization could take place.

Finally, the granting committee rejected the application because they assumed all the heavy hitters knew what they were doing. Yet, with the exception of Mark Davis (who was searching for the TCR in mice), they did not.

"They thought what they were looking for was part of the B cell," says Mak, they further assumed that the T-cell receptor was as abundant on the surface of T cells as it was on B cells. However, it turns out that, unlike the immunoglobulin-based receptor of B cells, which can be 10% of all the protein in B cells, the TCR comprises only 0.0001% of what's in a T cell. "And so, it's very difficult to isolate enough protein." They were expecting haystacks, while the Grail was just a few needles.

Regardless, there would be no grant. At that point, mentor McCulloch stepped in and more or less said, the hell with them. I will support you. Do the experiment.

And so, Mak did: "It took a year and a half to do: one postdoc, one technician, and me."

It Was a Beautiful Sunday

The experiment: Take all the genetic instructions to make a human B cell, and all the genetic instructions for a T cell, and swirl them together in a vessel, and then maintain that vessel at a constant temperature over an unspecified period while all that RNA and DNA mingles. Given enough time, the genetic strands that are similar in sequence will find each other in the mix and stick together. This is the intrinsic nature of singular genetic strands: They like to pair up. Subtracted from the mix are all those newly formed RNA/DNA couples. The remaining uncoupled genetic strands are, in this case, deemed unique to the makeup of human T cells.

Thus segregated, Mak and his team compared the unique T-cell sequences to other sequences from other cell types, as well as other species, where scientists already knew the protein's structure and related function. This last bit was enabled by something called GenBank, a DNA database curated by the NIH that collects DNA sequences from everywhere and anything. If a scientist wants to know what the protein encoded by his newly discovered DNA does—like all that uncoupled T-cell DNA of Mak's—one can probe the GenBank for similar entities. A match (or in science terms, an homology) can tell you a lot about the protein in question, even if the homology is weak.

The result: "It was a beautiful Sunday in June, 1983," says Mak. With the T-cell-specific hybridization and selection done, "we took those genes that were T-cell-specific and sequenced them. Not all of them were a complete sequence—in those days it wasn't easy to make a complete sequence—and I gave them to a summer student, Kathryn Leggett, an engineering student who wanted to spend time in a lab working on computers."

Leggett performed the search, probing GenBank with unique T-cell sequences. "There were 6000 sequences in GenBank at the time. Today, it's like 6 trillion sequences." Still, especially given the 90-pound weakling computers of the day, sifting through 6000 highly detailed bits of information involved a great deal of digital sweat.

"She must have done it all on Saturday because when I walked into my office on Sunday there was this two-foot tall pile of computer sheets on the floor . . . , all our bits and pieces of sequence, compared to everything that was in GenBank, just sitting there."

"I figured, it's a Sunday afternoon, nobody was there, and my wife had taken the kids to ballet or whatever, so I just started going through the sheets." After he scanned a few hundred pages, there was one that caught his eye, quite literally, because what he was looking at were not data as numbers, but as data-point patterns, visual repetitions of physical overlap. "I took the paper and held it up at an angle and I could see it." An apogee of overlaps. A match. "It wasn't easy to see because it was only a 5% homology, but it was so clear that over the V region and the D region and, I assumed, the J region, that it had that homology."

The match, designated clone YT35, was strongly suggestive of the critical molecular parts (V, D, and J regions) one needs for antigen recognition, and they matched with parts from a T cell. (For more on V, D, J, and antibody diversity, see the Glossary.)

"I knew at that moment that it was the T-cell receptor."

"I knew at that moment that it was the T-cell receptor."

The next day, Monday, Mak informed his little team. "I sat them down and I said, I think this is the clone that we should spend all the time studying. Forget about the rest. You may think I'm crazy, but I think this is the T-cell receptor," says Mak. "And of course, they thought I was crazy. We didn't tell anybody; I didn't even tell McCulloch."

Note: The investigators highlighted in this book often speak of being told they are crazy or of having crazy ideas. Since it is verboten to change direct quotes, should the reader tire of this word in quoted passages, please feel free to substitute "crazy" with any of the following: loopy, bonkers, nuts, wacky, daft, batty, mad, kooky, screwy, mental, cracked, feng (Chinese), and mishegas (Yiddish). Take your pick.

Team Mak spent the next five months verifying their results to be certain that, after all was said and sequenced, Tak Mak was not *feng*.

The paper announcing their discovery, a discovery at the heart of many of the technologies described in this book, was published in 1984 in *Nature*, and Mak's life hasn't been the same since.

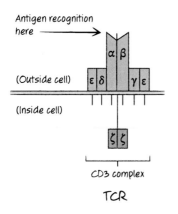

Note: The T cell receptor (TCR) complex consists of multiple similarly structured subunits (α, β, γ, δ, ε, and ζ—not unlike Legos—that have an emergent purpose when combined. Mak discovered the CD3 alpha (α) subunit, a revelation that led quickly to the discovery of the remaining subunits, including the so-called CD3 zeta (ζ) chain, the portion of the complex that passes the TCR signal on to the interior of the cell.

✤ ✤ ✤

What's the best thing about being a Canadian?

"Ten years ago there was an article in *Maclean's* magazine, which is our equal to *Time* magazine, and they did a survey asking who are the top 100 Canadians who ever lived. The number one of all time of the whole history of Canada was Tommy Douglas. He was a premier of Saskatchewan (1944–1961) and he said, I want universal healthcare for everybody in Saskatchewan, so no matter how old you are, how poor you are, when you are sick someone will look after you. Today, no politician would say anything to even touch on that or that kind of principle.

So why is Canada like this? I think it comes back to why you shouldn't go look for the T-cell receptor: Because big things, we should stay away from. We have our oil, we have our lumber, we have our gold . . . We live off the land and live a good life. In America and in other places, it is what you achieve, and if you haven't achieved, you're left behind. So somehow . . . I'm not stuck in Canada, but I just felt that that is a good starting principle."

ENGINEERING T CELLS TO ATTACK AND DESTROY TUMORS

Isolate T cells from blood of patient

Engineer T cells with TCR to recognize tumor

Engineer T cell to become an armed "super" killer T cell

Expand engineered cells to create an army

Infuse back into tumor-bearing patient

T cells seek and destroy tumor

Tumor

4/10/17

PHILIP GREENBERG
"Stealie Cells"

CHAPTER TWELVE

Philip Greenberg, M.D.

Head, Program in Immunology
Professor of Medicine and Immunotherapy
University of Washington
Seattle, Washington

ADOPTIVE CELL THERAPY (ACT)

"We were all arrogant enough to think that, yeah, this could be the beginning of something big." —P. Greenberg

First things first: When did you grow your hair out?
"In the 1960s. I actually haven't had a haircut since." His wife likes it. "She trims it occasionally."
Beatles, Stones, or The Dead?
"The Dead, of course. When we were in San Diego, Jerry Garcia used to come down to Encinitas and play. There was an old movie theater there and he used to play with the Jerry Garcia Band. Small place, maybe 150 people."
So, big Dead Head. "Plus, I used to show side-by-side pictures of Jerry and me. There's a likeness."
(You know, he's right.)
"I know I'm right."
Dead right. Until the data suggest otherwise.

"The Sky Was Yellow and the Sun Was Blue"

Philip Greenberg was born in 1946 in the Brooklyn neighborhood of Brownsville. "I enjoyed Brooklyn and I still enjoy going back, but it got to be a pretty rough place at the end," and Brownsville particularly so, even to this day.
"What I remember most about Brooklyn was taking the bus to go to watch the games at Ebbets Field," says Greenberg. Way before The

Dead there were the Dodgers, and as for all Dodgers fans, the move out of Brooklyn in 1957 was a massive hit. "That's when I realized how adults could disappoint you—it was an ugly reality—that you couldn't count on people."

The people that Greenberg *could* count on were his parents: solid, blue-collar folks. Mom was a secretary/administrator, and dad was a production manager in a factory. "Not a lot of science there in my background," says Greenberg, quickly adding, "But they did do the *New York Times* crossword puzzle every morning."

In pen?

"My dad, always in pen, yeah."

Knowledge was important. The need for a good education was strongly emphasized in the home, so much so that Greenberg was able to skip a couple of grades and enter college at the age of 16. That's the sort of thing you can do with smart parents, lax school regulations (at the time), and a couple of really good teachers.

The teachers were on Long Island, in the town of Baldwin, where Greenberg's parents moved for the duration of his high school years. "They thought I needed a different influence . . . they weren't crazy about my friends." It was in Baldwin High that Greenberg met his first academic mentors.

"There was Buzz Sawyers, a really remarkable biology teacher," recalls Greenberg. "The other teacher was the physics teacher, a guy named Stautt. And they were both just remarkable."

What made them stand out, especially Sawyers, was the ability to demonstrate and communicate passion. "If you can't show people that you're passionate and enjoy what you're doing, well, just transmitting information is not adequate." You have to care about how you're communicating; it isn't just telling people things, it's actually interacting. "That made them very influential as teachers."

And a good teacher can change the world.

"It's a curious thing," says Greenberg. In the years since high school, he's met two other investigators among his scientific peers who were also at Baldwin High at the time. "I didn't know them because they were a year or two behind me, but these were two women who ultimately became speaking faculty members at Stanford. They had exactly the same experience that I did." They went into science because of Sawyers.

"Goodbye Mama and Papa, Goodbye Jack and Jill"

For his undergraduate degree, Greenberg attended Washington University in St. Louis.

"We didn't have much money. I actually went to college on a Greyhound bus. My parents helped me pack my things and I got on a bus and went." He was 16 and headed for Missouri. Before that, the farthest west he'd ever been was New Jersey.

"I couldn't afford to live in the dorms; it was too expensive." He did manage to get shoehorned into an ad hoc housing situation, but it took getting used to. "I was in what were called 'faculty apartments' with three other people," says Greenberg. "It was kind of a strange mix. One of the other residents was a visiting professor from Korea; I remember that clearly because the first time I went to get something out of a drawer he had a 50-pound bag of rice in there."

The time in St. Louis was otherwise uneventful. Having earned his B.A. in Biology there in 1967, at the age of 21 Greenberg returned to New York for his medical degree at the State University of New York, Brooklyn. And yes, Woodstock happened during that time, but no, he didn't go. "People in my medical school did go, or at least tried to go. I mean, it wasn't easy to reach. You did get sort of frozen on the Thruway." There was a hippie jam.

After med school, Greenberg and his long hair headed (or perhaps migrated naturally) to the West Coast and clinical training at the University of California, San Diego (UCSD).

It was just the right groove.

"Ripple in Still Water"

So far as immunology goes, it was the right place and the right time. "It was a real hotbed," says Greenberg, some sort of weird convergence. UCSD had a recently formed medical school staffed by a top-shelf diaspora from the prestigious National Institutes of Health (NIH). Adding to that, a short drive up the coast in La Jolla was the world-renowned Scripps Research Institute. "It was entirely an immunology institute back then," says Greenberg, "And it was run by Frank Dixon [winner of the Lasker Prize in 1975], who recruited an incredible collection of people, virtually all of whom were experimental immunologists."

Within walking distance from Scripps was the Salk Institute, where one of the founding fellows, Mel Cohn (of the Melvin Cohn Award), had also put together a strong program in immunology. "So, again, you're surrounded by these people who are excited and can transmit that excitement."

The field was wide open: Being an immunology expert back then meant you were still half-clueless. "It didn't seem like people knew that much, so you could just jump in," says Greenberg. There were enough scientific questions to last a lifetime.

After he finished his residency, Greenberg made the decision to take several years to do basic immunology research, specifically in immunogenetics, the discipline of hunting down genes involved in the immune response. The work was fascinating. The work was profound. The work earned a Nobel Prize in Physiology or Medicine for three of its luminaries in 1980 (Baruj Benacerraf, Jean Dausset, and George Snell).

This was still the mid-1970s and Greenberg had yet to even decide on a clinical specialty, but what to choose? He certainly could do a lot of good in the field of rheumatology. Even autoimmune diseases in general would be a large canvas on which to work. Instead, Greenberg chose adventure; he chose what caught his eye. He chose hematology.

At that time, a man named E. Donnall Thomas was the head of oncology at the Fred Hutchinson Cancer Research Center in Seattle, where he was developing the technology of bone marrow transplantation for patients with hematologic cancers such as leukemia or lymphoma. "They were starting to get some interesting results," explains Greenberg. The procedure was rife with toxicity, but Thomas' team was getting clinical responses, and the data strongly indicated the immune system was key. "Ultimately, it became clear that the immune response was responsible for most of the cures."

> "Ultimately, it became clear that the immune response was responsible for most of the cures."

To Greenberg's mind, the work was the first convincing demonstration that an immune response could actually cure somebody with a malignancy. The technology was still in its infancy but it seemed obvious that it would one day grow up to be big and strong. "We were all arrogant enough to think that, yeah, this could be the beginning of something." Something huge.

Greenberg and Thomas started talking. "I told him I thought there was no reason why we should have to do bone marrow transplants," says Greenberg, recounting how he dove right in. "I said, we should be able to just move T cells [and only T cells] and target the tumor, and then you don't get all this bystander illness from graft-versus-host disease [GvHD]." It was GvHD that was producing the toxicity; it was GvHD that was killing patients. "Don was very single-minded about developing bone marrow transplantation, but he was perfectly happy with the idea." So much so, that he invited Greenberg to join him in Seattle.

Philip Greenberg, M.D. joined the Fred Hutchinson Cancer Research Center ("the Hutch"), and the Division of Oncology at the University of Washington in 1976. E. Donnall Thomas, M.D., bone marrow transplant pioneer and good sport, was awarded the Nobel Prize in Physiology or Medicine in 1990.

"From the Northwest Corner of a Brand New Crescent Moon"

Once settled in, Greenberg signed on to work with fellow Brooklyn native, Alexander Fefer, a founding member of the Clinical Research Division at the Hutch and a close colleague of Don Thomas. Fefer came to Seattle from the NIH, where he'd been developing animal models for T-cell therapy.

The models were splendidly designed and the approach was pretty straightforward: Fefer used a mouse that gets leukemia if you inject the (so-called) Friend murine leukemia virus into its abdomen. Fefer (later, Fefer and Greenberg) would inject the same virus just under the skin of a genetically near-identical healthy mouse. From previous observations, they knew this injection would act like a vaccine, immunizing the mouse against the virus rather than giving it leukemia. The T cells from the immunized mouse were then used to treat the leukemic mouse. In theory, the injected immunized T cells should recognize and attack cells that were infected with Friend virus (in this case, the leukemic cells).

In practice, it worked. Tumors in the mouse shrank. It was a home run.

"To be honest, I thought we would be in people doing something much earlier because it looked pretty effective back then," says Greenberg. Unfortunately, although the method was sound, they lacked a proper target. The mouse model worked because Greenberg knew exactly what the target was: the virus. He could select T cells from the healthy mouse that specifically recognized that target and use them as a treatment, just like a drug. But what target do you choose for human cancers?

"After almost 15 years in that mouse model we finally realized that we didn't really have a target yet that we needed to go after." But then they found one. They didn't have to look very far. It was all over the hospital.

At the time this mouse work was being performed, the other potentially fatal threat to bone marrow transplant patients (besides GvHD) was infection. After all, the whole point of the transplant was to wipe out the existing cancer-ridden immune system and replace it with a healthy new one. In the interim, as the new immune system takes hold, the patient is extremely vulnerable to infections of any kind. One such, and a major killer of transplant patients, was cytomegalovirus (CMV).

"Now there was a target we could go after," says Greenberg. "We can always recognize a viral antigen because that's actually what we were targeting in our leukemia model." Viruses are foreign; they are really easy for the immune system to see.

CMV is similar to the herpes simplex virus in that once you get it, you've got it for life. However, also as with the herpes simplex virus, just because you have it doesn't necessarily mean you have a problem. The viral DNA might just sit there. A person with herpes might not have an outbreak for many months, or even years. Similarly, if you're infected with CMV and immunologically intact, it shouldn't be a problem; you won't even know you have it. In fact, by the age of 40, roughly half of all adults are infected with CMV yet remain completely asymptomatic; their immune systems keep the virus in check. However, if your immune system is compromised (say, by a bone marrow transplant), the virus will flourish and, if left untreated, can kill you. In the early 1990s, there were no effective treatments for CMV.

"So we studied healthy people to find out what the immune response was that healthy people were getting that was targeting CMV," explains Greenberg, "And once we knew that, we decided we could essentially prevent people from getting CMV infection by giving them T cells that would be CMV-specific."

The project was headed by Stan Riddell, who was then a postdoc in Greenberg's lab. "Very talented guy, Stan. He really took that project and ran with it." The work was transformative. "We developed this technology for expanding lymphocytes—we called it 'rapid expansion protocol'—which is still pretty much the one that everybody uses now."

If a bone marrow transplant patient tested positive for CMV, a healthy compatible donor who was also CMV-positive would be sought out. Lymphocytes were then extracted from the donor's peripheral blood, the T cells filtered out, and CMV-specific T cells were separated from the mix and placed in a dish (a bag, actually) and subjected to the rapid expansion protocol. In just a few short days, thousands of T cells multiply to become billions.

"We knew that all the cells we were going to give would be CMV-specific so there wouldn't be any risk of getting graft-versus-host disease from these cells," Greenberg explains, and that was indeed the case. When these cells were administered there was no GvHD, and the treatment protected patients from reactivation of their latent CMV infections. "We gave it to patients to prevent them from getting infected and it worked," says Greenberg. "It was the first illustration that you could give a human antigen-specific T-cell clones and prevent disease."

"It was the first illustration that you could give a human antigen-specific T-cell clones and prevent disease."

The results of the study were published in 1992 in the journal *Science*. Fanfare followed, not the least of which was a feature article in the *New York Times*. One imagines Greenberg's parents reading it at the kitchen table before turning to the daily crossword.

The work was groundbreaking, and it laid the foundation for a branch of cancer immunotherapy known as adoptive T-cell transfer (ACT).

"Costs a Lot to Win, Even More to Lose"

Riddell, Greenberg, and their co-investigators made headlines, but they had not cured cancer. They had cured a viral infection, and the same was more or less true with the mouse model: The target was a virus, not cancer. Given that, it was logical that Riddell's next project targeted the viral headliner of the day: HIV.

"We did the same strategy we were doing in CMV," Greenberg says, but there was a tweak to the construct based on a clinical concern. If an HIV patient had a really robust immune response after receiving therapeutic cells, they might go into a type of shock. Specifically, the patient might experience what is now referred to as "cytokine storm." The "storm" is a condition of extreme immune response in which the immune system floods your body with potentially lethal inflammation agents, just as in cases of severe sepsis (see June, Chapter 16).

The potential for a storm was disconcerting enough to Greenberg that he decided to leverage the nascent technologies of genetic manipulation to install a "kill switch" on all the T cells he was going to use for HIV. Should the patient start to go south because of the injected T cells, Greenberg would flip the switch and all the genetically altered T cells would die. As with the rapid expansion protocol, the switch was a first-time innovation that has now, in concept if not in practice, been widely adopted.

The gene that Greenberg and Riddell genetically spliced into their T cells to serve as a switch encoded an enzyme called thymidine kinase, a molecular machine capable of activating the **prodrug**, ganciclovir. (Ironically, ganciclovir was the first effective treatment for CMV, approved two years after the conclusion of the CMV experiments.)

*A **prodrug** is a drug designed in such a way that it has to be activated by a chemical/enzymatic reaction in the body after the drug is administered. Such designs are used to reduce toxicity of a drug or to target the drug to certain tissues where the activating agents are present. Essentially, prodrugs are grenades that await the pulling of the pin.*

Greenberg's switch was a genetic replication within the T cells of CMV's vulnerability to ganciclovir. "So we put the viral thymidine kinase into these T cells and we gave it to these HIV-infected patients, with the idea that if they got this really potent inflammatory response, we could get rid of the cells."

He was being cautious, but it was unnecessary. "I guess the good news is that we saw HIV patients had fevers and they got mildly ill [from the treatment], so they did get a real inflammatory response, but not one that required ablating the cells."

The bad news was that the treatment didn't work.

Note: There were no donor cells in this experiment. The cells came from the patient. This was possible because HIV preferentially infects CD4 helper T cells, not CD8 killer T cells. It is this latter population that was filtered out from the patient's blood, genetically altered, expanded, and used for the treatment.

"What we found out very quickly was that while giving patients a second infusion of the cells, they rejected them," says Greenberg, still just a bit "I'll-be-damned" at the result. "This was quite interesting because these were HIV-infected people who had fewer than 50 CD4 cells—these were really immuno-incompetent people—so the magnitude of their immune response *to their own T cells* expressing one foreign protein was quite startling, because with the second infusion, cells typically lasted less than an hour."

What the results strongly suggested is that the cells, before they even had a chance to attack HIV, had instead acted like a vaccine. On injecting the cells, the patient's immune system reacted to the foreign thing they sensed: not the T cells, but the switch.

Although this work had nothing to do with cancer, this scientific aside is included here because there is no such thing as a failed well-designed experiment. Instead, the experiment told them that a T cell loaded with foreign antigen could act like a vaccine, and a powerful one at that. This observation is now being leveraged by Stan Riddell (still at the Hutch) in his work on cancer vaccines.

The observations from the CMV work were carried forward by Greenberg and crew and applied to skin cancer. To date, the antigen had been viral, and the specific molecular nature of that antigen had been previously characterized in detail. In contrast, skin cancer cells are not infected with a virus. Rather, parts of the cell are mutated, a much subtler difference as far as the immune system is concerned. To attack a cancer cell using ACT, you need to know the mutation.

"At that time, Thierry Boon's group in Brussels (see Gajewski, Chapter 24) had cloned an antigen that was specific in melanoma," says Greenberg, and that was Melan-A/MART-1. At roughly the same time, Steve Rosenberg's group at the NIH (see Chapter 13) independently cloned Melan-A/MART-1, as well as the melanoma antigen gp100. Both antigens were used by Cassian Yee, a postdoc in Greenberg's lab (now at M.D. Anderson), to identify melanoma-specific T cells and develop them as a treatment, using the techniques that had been refined for CMV.

It worked.

"In 2002 we published a paper in *PNAS* showing that we could give antigen-specific clones to a patient and potentially cure them of melanoma, which was pretty exciting."

> *Note: For any number of reasons this approach is not currently the standard of care for advanced melanoma (see Sections I and II). Nevertheless, Greenberg is still working on it. Investigations are also continuing for the use of T cells in combating CMV, HIV, and leukemia.*

❧ ❧ ❧

If it's not apparent by now, the cancer immunotherapy crowd is a family —as dysfunctional as all others—but still quite close. One of Phil Greenberg's closest friends, whom he met when they were both postdocs in San Diego, is James Allison (see Chapter 1).

"So, we were at a meeting in Hawaii, in Maui, and it turned out that Willie Nelson—he has a place in Maui—was doing a benefit for the Montessori school system. It was a black tie affair, so of course Jim says, 'Let's go crash it.'"

"We had a convertible, so we drove to the hotel where it was and we pull up—and we're dressed in jeans you know, pretty loose, and this was a black tie thing—we pull up and the guy who's parking the car says, 'Are you two in the band?' And we said, 'Yeah,' and he says, 'Okay,' and he takes the keys and tells us where to go. And so we just walk in."

"It was incredible. This was a black tie affair and we're obviously not in black tie, but there was an open bar so we just started drinking. We found a table and sat down. Nobody cared. And then Willie, him and his sister, they performed. After the performance, they came out and we spent some time with him. He actually acted like he recognized Jim, but honestly he was so stoned that there was no way he would recognize anybody."

Somebody took a picture of the three of them.

Greenberg still has it.

❧ ❧ ❧

Once in a while
You get shown the light
In the strangest of places
If you look at it right.

—Hunter/Garcia

Acknowledgments

Lyrics from "Scarlet Begonias": Words by Robert Hunter. Music by Jerry Garcia.
Copyright © 1974 ICE NINE PUBLISHING CO., INC. Copyright renewed.
All rights administered by UNIVERSAL MUSIC CORP.
All rights reserved. Used by permission.
Reprinted by Permission of Hal Leonard LLC.

Lyrics from "Ramble On Rose": Words by Robert Hunter. Music by Jerry Garcia.
Copyright © 1972 ICE NINE PUBLISHING CO., INC. Copyright renewed.
All rights administered by UNIVERSAL MUSIC CORP.
All rights reserved. Used by permission.
Reprinted by Permission of Hal Leonard LLC.

Lyrics from "Ripple": Words by Robert Hunter. Music by Jerry Garcia.
Copyright © 1970 ICE NINE PUBLISHING CO., INC. Copyright renewed.
All rights administered by UNIVERSAL MUSIC CORP.
All rights reserved. Used by permission.
Reprinted by Permission of Hal Leonard LLC.

Lyrics from "Terrapin Station": Words by Robert Hunter. Music by Jerry Garcia.
Copyright © 1977 ICE NINE PUBLISHING CO., INC. Copyright renewed.
All rights administered by UNIVERSAL MUSIC CORP.
All rights reserved. Used by permission.
Reprinted by Permission of Hal Leonard LLC.

Lyrics from "Deal": Words by Robert Hunter. Music by Jerry Garcia.
Copyright © 1977 ICE NINE PUBLISHING CO., INC. Copyright renewed.
All rights administered by UNIVERSAL MUSIC CORP.
All rights reserved. Used by permission.
Reprinted by Permission of Hal Leonard LLC.

CHAPTER THIRTEEN

Steven Rosenberg, M.D., Ph.D.

Chief, Surgery Branch
Head, Tumor Immunology Section
National Cancer Institute
National Institutes of Health
Washington, D.C.

INTERLEUKIN-2 (IL-2)

"When I began my work there was no such thing as a tumor antigen."
—S. ROSENBERG

Steve Rosenberg (no one calls him Steven) was born in the Bronx, New York in 1940. "Well, sort of the mid-Bronx, off the Grand Concourse, the part of the Bronx you might not want to go through today." Rosenberg's parents settled there after immigrating separately to the United States from Poland while in their teens.

"They never finished school," says Rosenberg, "But they were very smart. There was always a lot of questioning that was going on." The children, two boys and a girl, learned life by curiosity and by example: ask questions; work hard.

Rosenberg's interest in science came early and, again, by example.

"I always knew from a very early age—six or seven years old, when I stopped wanting to be a cowboy—that I was going to become a doctor and I was going to do research." His first mentors in that regard lived in the same apartment. "My older brother, he's 12 years older than I, he became a surgeon also." Both older brother and sister saw to it that young Rosenberg had the books he needed to stoke his budding interest in all things science.

Proximity of neighborhood also played a large role.

"I was lucky. I went to Bronx High School of Science, which is an amazing school—it takes students from all over the city—and that was the first time that I got challenged," the very first challenge being just to get in. "I had a large group of friends, but I was the only one that actually got into Science. I had to take exams and do things." Once there, Rosenberg quickly realized that

129

there were *loads* of smart kids in New York City, some even smarter than him, and furthermore, "You really had to work hard if you were going to get things done."

Note: Bronx High School of Science counts eight Nobel laureates among its alumni, which is, according to their website, "more than any other secondary school on the planet." Other graduates of slightly lesser distinction include astrophysicist and science communicator Neil deGrasse Tyson (1976), as well as Robert A. Moog (1952), inventor of the Moog synthesizer, featured prominently in the music of Stevie Wonder.

There was the supportive, inquisitive family and an amazing local school, but there also was history, recent history at that time, which was the prime mover for Rosenberg. "I was born in 1940," Rosenberg explains, "and by the end of the war most of my parent's families had been wiped out in the Holocaust." The news of it, of all the loss, arrived in the form of post-cards. "I remember at age six, postcard after postcard coming in informing us that so-and-so was murdered in Buchenwald or Auschwitz." Brother, cousin, aunt, niece . . . all gone. As this morbid mail piled up the little boy tried to make sense of it. "It was just a terrifying concept to me that people would be so evil towards one another, I mean, what people should do is try to help each other, right? So from the time I was six or seven years old I knew that's what I wanted to do." The most obvious choice was to become a doctor: to do research, to save lives. "It's the closest I can come to some explanation for why I've spent the last 41 years working six or seven days a week trying to get this done."

After the School of Science came undergrad and then medical school at Johns Hopkins in Baltimore, followed by a surgical residency at Brigham and Women's Hospital in Boston. "Then after my internship year I took off four years and got a Ph.D. in biophysics at Harvard."

Four years of hunkering down with the most convoluted of equations, just because.

Actually, biophysics was a canny choice because once you peel away the gooey layers of life, what you find underneath is math: the math that explains how your eye can focus light, or how liquid flowing in a tube creates blood pressure.

"I studied biophysics because at the time I just wanted to know as much as I could about everything," says Rosenberg. "And it's not a bad model for someone who wants to do things that are new . . . not just create variations on what's known, but take things in a new direction."

The extra coursework might seem like overkill to some, or the subject matter too daunting, but Rosenberg was adamant about hitting his ignorance head on. "In my view, education is losing your fear of what you don't understand." He wanted a wide enough base of understanding so that when he encountered future knotty unknowns, they presented as something that one could resolve, rather than an intractable tangle.

With biophysics out of the way, Rosenberg completed his residency in 1974. The next day, the leadership of the National Cancer Institute (NCI) appointed Rosenberg to the position of Chief of Surgery. It is a position he still holds.

Entree to IO

It was a curious thing.

"When I was a junior resident at the West Roxbury VA Hospital I saw a patient who came in with right upper quadrant pain," recalls Rosenberg. "He was having a suspected gallbladder attack." A cholecystogram confirmed that suspicion, and the course of treatment was determined: removal of the patient's gallbladder—a simple procedure. However, scarring on the patient's abdomen indicated he'd already had a far more invasive procedure. When asked, the man blithely recounted that he'd been operated on to remove a cancer. Years ago. Nothing special.

Rosenberg pulled his history. "His chart showed me that 12 years earlier he had been in the same hospital with a gastric cancer, a stomach cancer that had spread to his liver." The surgeon of record noted that he removed what tumor tissue he could, but that there were many inoperable metastases remaining, and therefore, the patient would eventually (probably soon) succumb to his disease. With no other therapeutic recourse, the hospital discharged the patient and sent him home to die.

"But then as I turned the page of his chart I read that he came back for a checkup three months later, and then again six months later, and then I read that a year later the man was back at work."

There were only two possible explanations for this: Either the original diagnosis was just plain wrong (unlikely, but possible) or the patient had experienced a spontaneous regression of his cancer—and this conclusion was so unlikely that it was considered to be impossible. There were, to that date, only four cases of spontaneous cancer remission in the medical literature. Four. Out of the millions of cancers diagnosed by modern medicine, there were exactly four.

Returning to the original lab reports and tissue samples, Rosenberg was able to dismiss the first possibility. That left the glaring truth of the second.

"He had undergone one of the most rare of events in medicine—a spontaneous regression of his cancer, and as a resident that substantially struck my fancy."

It seemed like the stuff of magic.

Hoping to tap into that magic, Rosenberg performed a very simple experiment: "There was another patient in the hospital at the same time with the same blood type, and who also had gastric cancer. I got permission to transfer blood from the patient who had undergone this spontaneous regression to the other patient, thinking that maybe there was something in the immune system of his blood that would cause this regression." It was a bold move, and one that typified Rosenberg's experimental efforts throughout his career.

> "He had undergone one of the most rare of events in medicine—a spontaneous regression of his cancer, and as a resident that substantially struck my fancy."

"Nothing happened, of course," says Rosenberg. The transfused patient went on to progress and die. "But that was the start of my interest in immunology and cancer." To stoke that interest, Rosenberg took a leave of absence from his surgical training and spent a year at Harvard delving into the mysteries of the immune system.

A Few Notes on an Otherwise Blank Slate

When Rosenberg was installed at the NCI in 1974, there were very few things known regarding the immune response and cancer. In humans, the work of William Coley, a surgeon practicing in New York in the early 1900s, very strongly suggested that the immune system could be awakened to the presence of a tumor by the introduction of certain types of bacterial infections. However, the specifics of this effect remained obscure because Coley was doing procedures, not controlled investigations.

Years later, work with animals done by a handful of researchers (the husband-and-wife team of Karl and Ingegerd Hellstrom, to name two) showed that they could induce an immune response to cancer in the laboratory. Further, they suggested that the immune response originated in T lymphocytes. It was elegant work, but immunologists were generally dismissive of their findings.

"You have to remember, when I began my work there was no such thing as a tumor antigen," says Rosenberg. "People didn't believe, in fact, that there was an immune response against human cancers." As late as 1957, the *Journal of Immunology* did not even list the word "lymphocyte" in its index.

Nevertheless, to Rosenberg the T-cell idea seemed entirely plausible. After all, in the setting of organ transplantation, it is the T cells of the immune system (not B cells, where antibodies come from) that are the targets of immunosuppressive drugs; T cells are in charge of destroying foreign cellular invaders. From there, Rosenberg merely distilled the concept of what could be considered "foreign." As far as he was concerned, cancer cells are different enough from healthy cells that the T cells of the immune system should be able to recognize and attack them. T cells became the focus of his work and remain so to this day.

Try, Try, Again

"When I started my work I did a lot of naïve things," Rosenberg admits. First, there was the hoped-for therapeutic transfusion of the mystery immune components described above. Later, after his arrival at the NIH, Rosenberg started to administer pig lymphocytes to cancer patients. It was already known at the time that if you inject an animal—say, a rabbit or a pig—with some type of antigen (like those present in tumor tissue), that animal's immune system will mount an attack against it. Given that observation, it was reasonable to assume that the parts of the immune system that performed the attack, whatever they might be, could be isolated from the bloodstream of that animal and used as a research tool or as a potential therapeutic.

Rosenberg's idea was to inject, and thereby immunize, a pig with a patient's tumor tissue, then extract the pig's lymphocytes—its T cells (which, theoretically, should be primed to recognize and attack any cell showing the antigens of the injected tumor)—and then administer those cells to the patient.

A great deal of preliminary work in mice sent a strong signal that such an approach would work in people.

It didn't.

"I immunized pigs against human tumors and transferred the pig cells into patients and nothing happened," admits Rosenberg. So much for strong signals. Two years of preparatory to work went down the drain.

That said, the work that went "down the drain" would not have even been permitted in today's highly regulated clinical research environment. Whether this development is good or bad for science is a matter of perspective. "When I started doing this work, the regulatory issues weren't nearly the same as they were today," says Rosenberg. "When I wanted to transfuse blood from one patient to another I just called the Chief of Surgery at the West Roxbury VA hospital, Brownie Wheeler, and say, 'I want to do this.' He said 'Fine, send me a few paragraphs', which I did, and we did [the experiment]." Then, Rosenberg wanted to give pig lymphocytes to humans—something that would be unthinkable today because of the discovery of potential adventitious viruses present in pig tissue—but at the time, this proposal underwent the same "wave-it-on-through" process. Write up a few pages, get a sign-off from NIH higher-ups, proceed. "I had a lot freer rein to try new things then," says Rosenberg.

This is not to say there was never any pushback to Rosenberg's clinical intents. There were, but the demurrals came from two very different camps. In camp one, Rosenberg's clinical colleagues thought that the whole pursuit was theoretically unsound. "They simply didn't feel that the immune system could be turned against cancer."

Of the second camp, the regulators, their objections were the product of sheer ignorance of the related science. "One of the first people [at the FDA] that regulated what I did had a Ph.D. in electron microscopy," says Rosenberg.

> *"One of the first people [at the FDA] that regulated what I did had a Ph.D. in electron microscopy."*

He was a very bright individual, clearly so, but he knew nothing about immunology. "So it took a mutual education with me learning what the FDA was about, and them learning what I was about"—experts getting each other up to speed.

Atomic IL-2

One of those early educational efforts on both sides was to comprehend the profound clinical activity of something called interleukin-2 (IL-2), an extremely powerful cellular signaling factor inadvertently discovered in human cell lines by Gallo and coworkers in 1976 (Morgan et al., *Science* 193: 1007 [1976]). Within a year of that discovery, a second group (Gillis and Smith, *Nature* 268: 154 [1977]) isolated the mouse version of IL-2, thereby giving scientists a way to make quantities of IL-2 independent of the mouse, which in turn enabled myriad mouse–T cell experiments because, as it turns out, IL-2 is a fertility treatment for T cells: Give a squirt of IL-2—get a bucketful of T cells.

Hoping to leverage this discovery, Rosenberg attempted to expand a starter batch of 10,000 mouse T cells by adding IL-2 to the dish. After just five days, those 10,000 cells grew to a population of more than 300,000. It was a remarkable proof of principle. IL-2, this research tool, this drug, this multiplier of T cells, was exactly what Rosenberg needed because of his *next* big idea for curing cancer with immunotherapy: Remove T cells from a cancer patient, grow the cells in a dish with IL-2 until they numbered a billion or so, and then give the T cells back to the patient. The underlying assumption was that, of all the cells removed initially from the patient, at least a few of them would recognize the cancer. If you could greatly multiply those chosen few, an effective attack on the tumor could be mounted. The cancer would be swarmed, overwhelmed, and killed. This thesis was not mere conjecture. Rosenberg and others had previously shown that if one cultured tumor cells from a patient in vitro and then exposed those cells to a sampling of the patient's own immune cells they would indeed kill cancer . . . in the dish.

Concurrent with these cellular investigations, the Rosenberg lab showed that IL-2 all by itself could attenuate tumor growth in a mouse. These two approaches—using cells grown with IL-2, and the administration of IL-2 directly—were then investigated in humans.

Ultimately, neither strategy worked. Worse, IL-2 proved to be incredibly toxic: Every patient treated with IL-2 wound up in the intensive care unit (ICU), and some of them nearly died from the treatment. Eventually, all of them died from their cancers, regardless of the method used to treat them.

Immunotherapy for cancer was not working.

Nevertheless, Rosenberg kept trying, and he kept after the FDA until they finally gave him permission to do the obvious, which was to combine the two experiments: Administer a billion dish-grown T cells to the patient *plus* concurrent, systemic administration of high-dose IL-2. This way, Rosenberg hypothesized, the host of T cells combined with the clout of the T-cell turbocharging IL-2 would surmount any possible resistance the tumor cells had to offer. It was not a "Hail Mary" pass, but it was close. Rosenberg was getting discouraged; patients were coming to the NIH, to him, for a miracle, and for years, he was unable to provide it.

In 1984, that changed.

"In 1984," says Rosenberg, "After treating 76 consecutive patients, giving them cells from pigs, from other patients, giving them IL-2—first naturally derived, and then recombinant IL-2—before we finally went to very-high-dose IL-2 combined with T cells . . . the first patient we treated that way, Linda Taylor, had a complete regression of her melanoma." And Linda didn't just get much better. Linda was cured. At the time of this writing, 32 years later, Linda Taylor remains alive and well.

The initial report was a bombshell. "Her picture wound up on the front page of newspapers back then." Rosenberg himself landed on the cover of *Newsweek*, and the journal article chronicling the IL-2 breakthrough is one of the most highly cited papers in medical literature. The title of the paper (Rosenberg et al., *New Engl J Med* 313: 1485 [1985]): "Observations on the systemic administration of autologous lymphokine-activated killer cells and recombinant interleukin-2 to patients with metastatic cancer," could have just as easily been called "Cancer Immunotherapy Works!"

A Reckoning on the Nightfall

How do you unwind? Got any hobbies?

"I don't have hobbies, and if somebody applies to work in my lab and they list all kinds of hobbies on their CV it goes into a separate pile. My wife will tell you that in the last 30 years there have been maybe 30 days when I have not been in the hospital when I've been in town. Otherwise, I go to the hospital every day. On weekends, I won't stay all day, but it's like what I tell our fellows: If you want to make progress, if you want to do new things, it requires two basic properties. The first thing is you have to be extraordinarily passionate about what you do. You have to immerse yourself, it has to be in your brain all the time, whether you're in your car and stopped at a red light, or when you're taking a shower, anything, you have to be thinking about the experiments you're doing and how to overcome them. You've got to always be thinking about the science.

The second [criterion] is you have to have a laser focus on your goal. Why? Well, when you're exploring new science interesting things crop up all the time . . . there's always some Brownian motion that can excuse a zigzag journey toward progress, but unless you keep your eyes on the goal, unless you have a laser focus, you're not going to be the person who makes these new advances.

So, how do I unwind? It's much harder to unwind when you're taking care of cancer patients. I always have a dozen patients in the hospital, all of them dying of cancer, and they've come to us as a last hope. Some people refer to the NIH as the National Institute of Hope, and I wish I could tell you that when I lay awake at night I think of the successes in patients who have been cured, but that's not what I think of. I think of the patients that we did not cure, the patients whom we failed. So, you can't let yourself get away from it too much—maybe a little bit, sometimes—but not too far away or you lose that immersion that's required to make progress."

✤ ✤ ✤

Three final things:

- For some readers, especially those with a science or immunology back-ground, the story of IL-2 as related herein suffers from too short a telling.

This is a valid criticism. The story of IL-2 is fascinating, dramatic, and long enough for its own book. Indeed, such a book already exists: *The Transformed Cell: Unlocking the Mysteries of Cancer*, by Steven A. Rosenberg (Putnam Press, 1992).

It's a terrific read.

- The work of Steve Rosenberg did not end with IL-2. Shortly thereafter, with the trainees in his lab at his disposal, Rosenberg explored the potential of tumor-infiltrating lymphocytes (see Hwu, Chapter 15), checkpoint inhibitors (see Topalian, Chapter 6), CAR-T cells (see Eshhar, Chapter 14), and numerous other avenues of investigation (see Zitvogel, Chapter 25). All the highly accomplished investigators mentioned here, and innumerable others, trained in the laboratories of Steve Rosenberg.

  Presently, Rosenberg's laser focus is on something called tumor neoantigens, as investigated via next-generation DNA sequencing, and the probing of molecular libraries with tandem minigenes.

- To say that Steve Rosenberg, chief of the surgery branch at the NIH for the last 40 years, is a person of great scientific and political power and influence is a gross understatement. In closing, two examples.

  "One does not criticize Steve in public."   —ANON

And this:

"A colleague of mine asked me to give a lecture on the CAR T-cells, on the double chain. I was giving a lecture and Steve Rosenberg was there. Afterward, he was holding me at the door, he says, 'Zelig, what are your plans?' I told him, 'You know, Jeff made me an offer.' He says to me, 'You are staying here.' I say, 'Steve, I'm committed. We already discussed the project.' 'No, you are staying here.' [Big smile.] And so, I stayed. He brought me into the program and gave me all the facilities and all the help I needed. He gave me half a floor. All the other fellows were envying me. Since then we are very good friends."

—ZELIG ESHHAR (SEE CHAPTER 14)

SECTION VI

CHIMERIC ANTIGEN RECEPTOR—T CELLS (CAR-Ts)

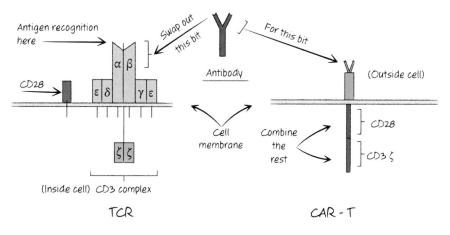

To be clear, the "swapping out" of bits as shown in the above diagram is actually done at the genetic level—it is not a literal cut-and-paste of protein bits but rather a cut-and-paste of bits of genetic blueprint that results in the manufacture of new "recombined" proteins, such as a CAR.

ZELIG ESHHAR
"Still Life 'Mouse with T-Body'"

CHAPTER FOURTEEN

Zelig Eshhar, Ph.D.

Professor of Chemical and Cellular Immunology
Weizmann Institute of Science
Rehovot, Israel

"We put sacks full of sand in the windows to avoid snipers."
—Z. Eshhar

Zelig Eshhar was born in 1941 in a small town called Rehovot, in a place officially designated as the British Mandate.

"It was a time of wars, before the state of Israel," says Eshhar, and this fact is reflected in his Israeli identity number. "My ID is just over half a million. I was citizen five hundred thousand something."

His parents were staunch Zionists who emigrated from Poland before the war. "My parents were what we call pioneers," says Eshhar, and they settled in a place that could not have been better for a budding naturalist: there were trees, there were hills, and there were critters.

"I never forget what my mother told to me once, she told me, 'Zelig, I knew that you [would] be a scientist.' I asked her, 'Why?' She told me, 'All the free time you had, ... you were on the hills around the house.' And what was I doing? I was collecting turtles, and snails, and looking on birds." He even conceived his first experimental design as a boy walking those hills.

"I had questions," says Eshhar. "For example, there is a bird that makes the nest like this." (He cups his hands together, leaving a hole formed by thumb and forefinger. "I don't know the name in English ... but this nest, it hangs from a branch and there is a small opening where the bird can enter." Young Eshhar reasoned that the nest's placement and its tiny entrance served to discourage predators.

What he couldn't figure out is how it got that way. "How does a bird, a young bird know how to do this?" Is it learned behavior? Eshhar had never witnessed any bird parents giving nest-building lessons to their offspring, "So, I said to myself, if I would take eggs of this bird and put it under a chicken until they hatch [the Eshhar family kept chickens], will the offspring know to build the nest?" (For the record, he did not execute this particular experiment and does not remember why—perhaps his grant application was rejected? We will never know.)

There were, of course, other influences. HUGE influences. Yes, trees, hills, and critters, but also, yes—Chaim Weizmann. As far as fate goes, that's almost cheating: Chaim Weizmann was a famous scientist, the first president of Israel, and Eshhar's boyhood neighbor.

"He was a chemist. He worked in England and he developed a kind of fermentation technology to make gunpowder that will not make smoke." The innovation was critical to the British war effort in World War I, and it made Weizmann a wealthy man.

"His house was in Rehovot," Eshhar recalls, "A very nice mansion surrounded in orchards. As a kid I remember walking around his house—a big place—and he had gazelles and peacocks and other things, and for us it was like having our own zoo."

Observing the life of a scientist made a big impression on a little boy, which, in part, led to him becoming a scientist: a famous scientist at the Weizmann Institute, the world-renowned research institution founded on the very grounds of the big house with the gazelles.

King of the Bees

Although firmly on the science path, there were some significant detours, the first being two years of state-mandated military service. Following this, Eshhar went to a kibbutz, a collective agricultural community called Yad Mordechai, named after Mordechai Anielewicz, one of the leaders of the Warsaw ghetto uprising during World War II. Yad Mordechai is located within walking distance of the Gaza Strip.

"In the kibbutz, each of us had to choose a discipline in agriculture," says Eshhar. He chose bees because (a) beekeeping is critical to agriculture, and (b) you can't tame bees. "You can't tame them! They do whatever they want. You are just there to serve them, to live with them, and wake up in the morning and put your nose in the air and smell where flowers are blooming and take the hive there." It was appealing because bees required that you understand them.

"I started to read books about bees. I learned so much about bees that even today I think I'm an expert on bees." It was his first passion. "I liked it. I wanted to learn more. [At] the end of the day I found books about beekeeping that told me I knew more than the experts."

But kibbutz activities were not all bees. While Eshhar was there, he attended a presentation by a guest lecturer, a scientist. "He gave a talk on this golden era of biochemistry when they found DNA, RNA, protein, all these things." Eshhar was fascinated and immediately informed his fellow kibbutz members that he was applying to university.

They were not happy. They told him to wait. They needed him to stay. "I didn't feel I [wanted] to wait." Eshhar had contacts in the faculty of the Agriculture Department of Hebrew University (founded by Weizmann) and managed to be accepted into the School of Agriculture, which just happens to be located in Rehovot, Eshhar's hometown.

While he waited for the school year to begin, Eshhar resumed his study of bees on a more sophisticated level. "I worked with a group of scientists doing fieldwork in the kibbutz. That's when I started to experience real work as a biochemist. I operated on bees, took different fluids from them … It was incredible. And I was good at it."

But every scientist has his setbacks. "One thing that happened to me, which was traumatic, is that my mentor decided to study wasps. So I told him, 'You know, let's go to Yad Mordechai, to the kibbutz … I know where there are wasps because wasps are eating bees and they are our enemies.'"

They went to the kibbutz, they sedated the wasps, collected them, and returned to the lab in Rehovot. Everything went fine until unexpectedly, all the wasps woke up. "And who did they go after? They attacked *me*. I came home … my mother saw me … She thought I was in a fight. I had a head like this." (The size of a basketball.) "But I survived."

Eshhar had had a nasty immune response. From that experience and others, he wanted to know why. He switched his study from biology to biochemistry to, eventually, immunology. The decision took him to Jerusalem.

War and Independence

Another academic detour intervened, this one called the Six-Day War.

Eshhar was trained as a paratrooper. "Right after starting my Master's at our laboratory in Jerusalem we were informed of the war. There was the Arab city of Jerusalem, and here we were. We put sacks full of sand on the windows to avoid the snipers." Shortly thereafter, Eshhar was called to serve.

"I was fighting in the Golan Heights. Not much. I don't want to tell stories, but we had prisoners. I gave them water. When I was told to clean the latrine, I did it. During night, we slept in the abandoned trenches of the Syrians because we were dropped by helicopter and had little equipment. So, we felt alone, but not alone. You had to depend on yourself."

In recollecting his war experiences, Dr. Eshhar speaks slowly, pausing often, and describes what seems to be the smallest fraction of what he's seeing in his mind.

"After the war my spouse asks me, 'Zelig, did you kill Arabs?' I say to her, 'I don't know. I hope not.'"

10, 10

Eshhar finished his Master's and was considering his options. "I had a few offers to stay in Jerusalem but then I am getting a phone call from a friend of mine at the Weizmann Institute. He says somebody is looking for a Ph.D. student. I say, 'Who knows me?' He tells me, 'Michael Sela.'" Dr. Sela was an immunologist, and one of the founders of the field of immunochemistry. "I go there. I say, 'Hi Michael, I'm Zelig!'" He was 28 years old.

Sela put him right to work writing a Ph.D. thesis proposal. "I wrote it as Michael suggests to me, and the committee reads it and then they tell me, oh, it's too presumptuous." Too much. Too ambitious. They say rewrite it. Don't aim so high. "So, what do I do? I rewrite. Now they say, okay, but by the end of my Ph.D., I did exactly what I originally suggested and even more."

The topic was his first deep dive into immunology and two critical components of the immune system: B cells and T cells. "My task was to prepare antilymphocytic serum, which is antisera against T cells, in order to suppress T cells so that they will not reject grafts, foreign grafts. That was my job."

It was an ambitious project for a Ph.D. student—to figure out how to prevent transplant rejection—a serious clinical challenge to this day. "But I succeeded. I vaccinated horses and goats. I took apart T cells." He immersed himself in the *system*, just like he did with the bees.

He finished the project and wrote it up. "Michael, says, 'Show me.' I left it there. The day after he say, 'Excellent.' I told him, 'You want your name on it?' [It's standard practice for the principal investigator (PI), the mentor, to be listed as a coauthor on publications derived from work leading to a Ph.D. thesis.] He said, 'No, go ahead by yourself,' because he didn't know what I did! So we are very good friends to this day."

With his Ph.D. finished, it was time for a postdoc. "Sela asked me, 'Young man, where do you want to go for a postdoctorate?' I told him, 'I know that in the U.S. someone is studying how to fuse cells together [**hybridomas**] and learn what is dominant and what is not.'"

Note: *Hybridoma cells,* alluded to here, are a method of producing large batches of identical antibodies—monoclonal antibodies—that have multiple investigative and therapeutic applications. The inventors, Niels K. Jerne, César Milstein, and Georges Köhler, received a Nobel Prize in Physiology or Medicine for this work in 1984.

"Sela looks at me and told me, 'Zelig, you have three kids (by then, I was married with three kids). You don't go with kids to New York City.' But he told me, 'You know what? I have a very good idea.'"

Sela put a call in to a friend at Harvard and lined up a laboratory position for Eshhar. The friend was the head of pathology, Baruj Benacerraf (Nobel laureate, 1980). "So I went to Boston without knowing barely any English. For example, people ask me, 'How are you?' So, I would tell them all about me and how I was. I didn't know. Honest to God, I was so naïve."

Eshhar spent three years in Boston dissecting T cells of all kinds, defining each discrete bit by its molecular weight. "I was involved with another project, also in autoimmune disease, so it was cancer antibodies, autoimmune diseases, immunosuppression, allergy ... I was touching everything. And I was apparently so good that after I left Boston and came back to the Weizmann Institute, Benacerraf called me." (There's a big smile and a pause here.) "He says, 'Zelig, how [did you do] this and this experiment?'" His performance at Harvard had earned Eshhar a reputation as a master technician, and his biggest technical coup was about to take wing.

"Later on people ask me, 'How did you figure this out?' Well, I read a book." The book was *A Feeling for the Organism*, by famed geneticist, Barbara McClintock. "She was looking on corn grains, which are black or brown, or some color, and she could define the rules of genetics just by looking on the corn." It opened his eyes to what he already knew (Eshhar is an expert in this sort of holistic observation). "So I start to think to myself, T cells, like with the beehive ... You know, everything was in my head. I could close my eyes and see it."

"When I worked with T cells I was thinking like it's a living organism that you study, and you give everything, all the parts, a name or function and then you know how the system works." Then you test the system, you perturb it, give a little kick in some way and see if the system responds as predicted, or not. That's how you learn, how you add

> *"You even dream on it at night, you continue to think about the system."*

another small part to the comprehension of the whole. "You even dream on it at night, you continue to think about the system. So if you ask me how I came to this idea ... I don't know, really. But it came to my mind."

What came to Eshhar's mind was something he called a T-body, an exotic new way of fighting cancer by directly bioengineering T cells.

Model T-Body: The First-Ever CAR-T Cell

The T-body was a radical idea, and served as a proof of principle, a prototype if you will, for an invention called a CAR-T cell (chimeric antigen receptor T cell), a cancer-fighting technology now the focus of a growing list of biotechnology companies engaged in a multibillion dollar drag race to build the first FDA-approved CAR-T.

This is how it works: Each T cell in your body is outfitted with a T-cell receptor (TCR; see Mak, Chapter 11). The receptor is there to identify and destroy anything that's foreign (i.e., that displays a given discrete target, referred to as an antigen). For example, when flu virus infects cells in your body, turning them into little flu factories, a properly instructed T cell can see those infected cells and put a stop to it. These "proper instructions" essentially amount to a molecular recognition pattern imprinted in the cell. This can occur by having seen—and successfully eradicated—a particular foreign infectious agent previously, or by having examples of what to look for provided by, say, a yearly flu shot. Regardless, this training—this T-cell memory of what is bad for you—is very effective and can last a lifetime.

Why, then, do people die of cancer? Because cancer isn't foreign. Cancer is part of us, and on a molecular level, the differences between healthy and cancerous cell types are hard for the immune system to distinguish. Also (as discussed in other chapters), cancer fights back.

While Eshhar explored the building blocks of the TCR—the part of the T cell that tells it what to attack—it occurred to him that the TCR was similar in many ways to how antibodies are constructed and function. Antibodies, made by B cells in the immune system, are also designed to pursue one distinct aberration, one antigen, to which they latch on and flag for destruction. For example, the purpose of a flu shot is not only to prime a T-cell response but also to induce the manufacture of flu-specific antibodies by B cells.

The main structural difference between antibodies and TCRs is that B cells produce antibodies and secrete them, whereas TCRs are firmly integrated into the cell membrane of T cells.

"It's so simple, you know." (Eshhar habitually says this about things that are appallingly complex.) "TCR recognizes antigen. There. So I said, antibody and TCR both belong to the same gene family, the same subunits, [and the] same rules. I'm thinking, it's so simple. Let's replace the variable region of the T cell with the antibody." This would be a much more accurate targeting system for any chosen adversary. The result, circa 1985, was the T-body.

The initial motivation of this work was simply to see if the destructive activity of T cells could be redirected. The idea of directing T cells to cancer came later. "The aha moment came because I knew you can use antibodies against a tumor," says Eshhar. Indeed,

> *"The aha moment came because I knew you can use antibodies against a tumor."*

there are a number of antibody-based drugs currently on the market that do just that (e.g., rituximab, an exceptionally effective drug used to treat certain blood cancers). Eshhar then realized that he might be able to overcome a TCR's inability to recognize antigens on cancer cells by replacing TCR parts with other parts—antibody parts—that do. Essentially, lop off the ineffective business end of the TCR and replace it with a tumor-antigen–targeting antibody.

This was some sophisticated science involving the genetic engineering of the TCR, which, in 1985, was a headache to even think about, unless you're Zelig Eshhar. "Because we knew all the components … It was kind of like a Lego. Snap them together. It was simple."

Enter the Mentor

"My mentor was Michael Sela. He could get excited about a good project. If you are good, he will let you have all the freedom in the world. I didn't even know what I was doing, but he let me. So that's a good mentor."

But it wasn't entirely smooth sailing, mentor-wise. Eshhar's experience with his PI at Harvard, Benacerraf, was not so positive—contentious, even: a matter of perpetual difference of opinion that, out of professional respect, will not be further described. Let's just say, it was a bad fit. Eshhar implemented work-arounds as needed, and then left as soon as possible.

Yet another mentor caused Eshhar undue grief, but he profited from it, as one does by sinking or swimming. The advisor in question (who shall remain unnamed, again, out of professional courtesy) led to Dr. Eshhar's Dark Night, triggered by the mentor leaving for a sabbatical in the middle of Eshhar's training. "And I said, 'Oh God, how I will continue?' Because I didn't know, I didn't think that [I was] good at this or not … Now I don't have a backup." Somehow, Eshhar found a way to use his panic. He scrambled. He attacked his project by mastering a method. The project was characterization of polyphosphate granules, and the method was electron microscopy, a way to image things that are really, really, *really* small.

"So I found somebody with electron microscope." Eshhar then studied how to identify polyphosphate, and the result was he learned how to operate the electron microscope. "At the same time, I became friends with everyone in the lab and I was asking questions, and they supported me. When I realized my mentor [was] not coming back, I didn't care. I didn't care because I realized in retrospect that it taught me to be independent."

✤ ✤ ✤

Endnote: There is no approved drug to mention here. Eshhar's work in CAR-Ts was a critical proof of principle, but it was not clinically effective. Even Legos can be hard to work with if you don't have all the right pieces. Eshhar was still missing a piece. (See June, Chapter 16.)

Introduction of chemokine Receptor genes into T-cells to improve their migration to tumor

chemokine Receptor

T-cell

Improved migration to tumor

chemokines

Tumor

5/20/16

Patrick Hwu

PATRICK HWU
"Tweaking the T"

CHAPTER FIFTEEN

Patrick Hwu, M.D.

Head, Division of Cancer Medicine
The University of Texas M.D. Anderson Cancer Center
Houston, Texas

"Why don't you just inject dirt?"
— P. HWU, QUOTING A CHEMOTHERAPIST COLLEAGUE

Patrick Hwu was born in 1963 in the town of St. Albans, a place no one has ever heard of near the metropolis of Charleston, a city itself consistently misplaced because it is *not* the one you're thinking of in South Carolina, but is in fact a city in West Virginia, a state that itself is a locale of some existential uncertainty.

"No one even knows West Virginia is a state," says Hwu, "They think it's the same as Virginia, except western Virginia, and they don't know Charleston is a city there," even though it's the state's capital. And St. Albans? Forget about it. "So, let's say I'm from a very obscure place in southern West Virginia."

✤ ✤ ✤

Patrick Hwu, M.D., is from a very obscure place in southern West Virginia. This was something of a singular experience in his youth, made slightly more so by his ethnicity. "It was a great place to grow up, but yeah, I was probably the only Asian guy in the state at the time."

So, what did the Asian kid want to be when he grew up?

"In high school I was thinking about law, or journalism, or medicine," says Hwu, relaxed, sitting back, recalling his near-miss careers with his perpetual, weary cheer. "I was Editor of my high school paper, and I actually did a little internship with *Charleston Gazette* too. I got to write all the really insignificant, short articles." This trial-by-fire stint in daily news journalism resulted in Hwu's learning of the Associated Press writing style, and his earning of a handful of those delicious bits of ego food called clips: stories with his byline.

"Yeah, little clips." Writers adore such things, but as compelling as these scattershot validations were, Hwu was uncertain how best to use his burgeoning skill set. He discussed his future with his fellow guardians of the Fourth Estate and they were all very supportive. "I told the other reporters I was thinking about pursuing journalism or medicine, and they were like, 'NOT journalism!' And I mean, it was unanimous. So I became a doctor."

Dr. Hwu

The decision to become a doctor took Hwu to Lehigh University, followed by the Medical College of Pennsylvania, then to Johns Hopkins for a residency, and finally to the National Cancer Institute (NCI).

"I short-tracked in the NCI where I did my fellowship in oncology," says Hwu, although while at the NCI he seriously considered becoming a pure Ph.D.; his interests were primarily in doing research. "But I decided to get my medical degree because then you could work in a lab, and see patients," and he already knew he wanted to do cancer immunology.

"I just thought it was an interesting idea that should work," says Hwu. After all, our immune system fights off most other diseases, and often far more efficiently than pharmaceuticals. "In fact, if you do the calculation we've probably prevented more occurrence of infectious disease and more death by using vaccines [an immunotherapy] than we have using all those antibiotics."

The best way to address his interests at the time was to train with one of the reigning masters in the immuno-oncology field, Steve Rosenberg (Chapter 13) at the NCI. Rosenberg's extensive track record in IO suggested he would be an appropriate mentor; with Rosenberg as a guide, and the immune system as a tool, why not cure cancer?

Enter the Mentor

The two names that appear most frequently in the background of IO researchers are the late Lloyd Old and the still-very-much-with-us Steve Rosenberg, Chief of the Surgery branch of the NCI. "Steve built an incredible team of people focused on trying to attack cancer with the immune system. Everything he did was focused on that goal," says Hwu, and it was a powerful example.

The goal was curing cancer; the prestige of the lab and the continuing career advancement of the investigators training there were incidental. Stringent resource management was key in achieving that goal. "There would be interpersonal issues here or there, or other [distractions], but Steve just kept everyone thinking about the goal," recalls Hwu, and he did this in part by embracing both the strengths and weaknesses of his team. "He got such an eclectic group of people to work together and have each person realize their particular skill set to get that done." In this way, the individual excelled and thereby the group, but the benefit was all for the science.

Even when it came to reporting on the lab's discoveries, the goal was still the priority, which was something of a risk in the scientific world. Since the pressure to get your work published is huge. The right paper in the right journal could earn you tenure or ensure your next grant. "He had a very high bar for the science," says Hwu, "It's very clear. Steve was always making sure things were extremely correct before they went out, even at the expense of being first."

Lessons learned at the NCI play out today in Hwu's lab at M.D. Anderson. First and foremost is keeping one's eyes on the prize, while leveraging the advantages of playing well with others. There was also something about not having doors: "Steve never had individual labs where you go in there and you lock the door and say, 'This is my lab and this is my lab's equipment.' So, like Steve, what we have is what we call 'the big lab,' a lab where everyone works together, everyone shares things."

The sharing of knowledge and investigative tools is a core principle in the Hwu lab, a standard that requires constant reinforcement. "I had a talk about it this morning with the group," says Hwu. A postdoc wanted to present at a scientific conference but was reluctant to include some critical details of the work for fear of being usurped by a copycat. "And I said to him, I said to everyone, my whole group, I said, 'You know, let it out there, get it out there. What a wonderful thing if someone scoops you and cures cancer first. And if someone does scoop you after we have such a huge lead, shame on us, right?' Don't be protective. Get it out there. That's one of the key lessons I learned from Steve."

> "What a wonderful thing if someone scoops you and cures cancer first."

Still, it can be a hard argument to make if you can't maintain the broader philosophy of "If you give it, you get it back." "If you're open, the pie gets bigger, everyone has enough stuff," says Hwu. In fact, the scientists that keep a secret for fear of being scooped, "those are the people that aren't really helping society as much."

What about getting grants? Don't you need to protect your work for that? Hwu's experience has taught him that the people that are most open with their findings, their technology, are actually more competitive for grants. "They have many more interactions, more interfaces." By default, the open collaborative network casts a wider net.

For example, consider the so-called pmel mouse, an experimental critter developed at the NCI by Nick Restifo (still at the NCI) and Willem Overwijk (now at M.D. Anderson). The mouse has a very special immune system capable of recognizing human melanoma tumors. Restifo and Overwijk could have kept their unique (and extremely useful) investigative tool in-house if they wanted—they were under no obligation to share—yet they distributed it freely. "They were giving that mouse out right and left," says Hwu, "And that was at least two years before their first paper came out [using the new pmel mouse model]. It didn't hurt them at all. In fact, it helped them because they gave out so many that now it's the standard mouse model that almost everybody uses."

Hwu doesn't just preach this approach, he practices it. "When we published [our group's] IL-21 cancer paper around the year 2000 or so, I sent the plasmid out to anyone that wanted it. My postdocs wanted to kill me because they were almost scooped. It turned out fine, but again, if we did get scooped, shame on us. So send it out to people, work with people, share. This helps society move faster. Remember, it's cancer that's the scourge."

TILs, Then Aha!

Hwu landed in Steve Rosenberg's lab in 1989, when they were working on a drug called IL-2. However, cell-based therapies were also coming into focus and part of that work revolved around tumor-infiltrating lymphocytes (TILs), which are immune system cells that have infiltrated the tumor mass, presumably to attack it. The problem is that once these lymphocytes have infiltrated, the tumor cells send out signals (such as PD-L1, IL-10, and others) that tell the T cells to stop the attack and just sit there. This, they do, disastrously so, and this is where Rosenberg, Hwu, and others found them, extracted them, and began to toy with them to get them working again.

Briefly, the therapeutic program involved removing the stymied TILs from the cancer patient, innervating them in some way, multiplying them in a dish until there are billions of them, and then shooting the whole lot back into the patient. The idea was to physically overwhelm the negative signaling coming from the tumor.

The first step was to work out the basics: keeping the TILs alive. "When I first came on board I worked on a project looking to mark these T cells to see how long they survive." This is something you need to know if you're going to use T cells as a therapeutic, because although pills don't get tired or die, T cells do. To detect living T cells in a sample, Hwu inserted a gene into the T cells that encodes an enzyme called neomycin phosphotransferase.

This new gene protected the transfected cells from the lethal effects of the drug, neomycin. The next experiment consisted of injecting the transduced TIL cells into the subject (be it human patient or mouse), wait days or weeks, take a sample of cells from the test subject, give that sample a spritz of neomycin, and then look for T cells that are not dead. "Guess how long the T cells lasted in the body?" prompts Hwu. "Three weeks." That's it. Three weeks, instead of years, like regular T cells. No wonder TIL therapies weren't working: The transferred T cells were dead before they ever got going.

The mechanistic reasons for their failure were several-fold but in the most general terms, the niches where lymphocytes reside were too crowded, and the crowd, on the whole, was unfriendly. "We're still not even sure if it's a real physical space issue, or something else," says Hwu.

To improve the chances of TILs surviving long enough to function, Hwu and colleagues borrowed techniques from the world of stem-cell transplantation. Such conditioning regimens are designed to destroy the patient's existing immune system from which a cancer has arisen (e.g., lymphomas, leukemias) before replacing it with the healthy stem cells that will eventually reconstitute the patient's entire immune system.

One such lymphocyte-depleting regimen used the chemotherapy drugs, fludarabine and cyclophosphamide (flu/cy). In theory, borrowing this method for use with TILs seemed reasonable, but Hwu had serious reservations. "I was the only medical oncologist in the group—the others were surgeons—so I was the only one who really gave a lot of chemotherapy in the past and I was thinking, this is going to be really scary, it's going to be very toxic ... but we've got to try it." His patients were dying. They had to do something.

"But it turns out that flu/cy was well tolerated," admits Hwu. In fact, the TIL treatment, which included the growth factor called IL-2, a booster for T cells with its own serious toxicity issues, was less toxic when patients received the flu/cy lymphodepletion. "We understand now that we probably depleted a lot of the CD4 cells that were the actual source of the toxicity." Hwu theorizes that the IL-2 growth factor dangerously supercharged the activity of CD4 cells, transforming them from disciplined guard dogs to rabid pit bulls. Further, the lymphodepletion protocol eliminated cell types with T-cell-suppressive activities: some called regulatory T cells, or Tregs (see Sakaguchi, Chapter 20), and others called myeloid-derived suppressor cells, or MDSCs (see Gabrilovich, Chapter 23).

"The patients started doing really well," and Hwu had to adjust his thinking. "Initially with TILs I was like, wow, this isn't going to work because even though it was working great for about three weeks, tumors would recur [after the TILs died]. But then we did the flu/cy, and that was the aha moment.

We really got very excited about TIL therapy after that," says Hwu, adding, "To this day, lymphodepletion is being used in every adoptive cell therapy protocol."

CAR Talk

TIL adoptive cell therapy is still very much under study, but at the time Hwu was performing the above investigations, instituting TIL therapy was no easy task. TILs were hard to grow, hard to keep alive, and (at the time) not amenable to the treatment of tumor types beyond melanoma. "We knew that T-cell therapy worked in melanoma. The question was, why can't we get that therapy to work in other cancers?"

Hwu's work with TILs greatly facilitated a solution to the next step in T-cell technology: the chimeric antigen receptor T cell, or CAR-T. (A chimera is something that has organic parts from at least two different sources. Pegasus, the horse with wings from Greek mythology, is a chimera.) "My initial project as a fellow was to put the TNF [**tumor necrosis factor**] gene into TIL cells in order to get the TILs to first migrate, and then secrete the factor right at the tumor site," explains Hwu, but progress was slow. Lymphocytes don't like to express foreign genes as well as other cell types, and high-tech options to boost the success rate of this process were few. "This was a long time ago, so we didn't have great **vector** systems or anything. It was really hard to transduce a lymphocyte back then."

Vectors: Think of vectors as a biological shuttle bus where the passenger is the set of instructions (DNA or RNA) that makes up a gene, and the drop-off point is another cell where those newly delivered genes are put to use. At the time of Hwu's initial experiments, the only shuttles available for gene transport were as reliable as borrowing your friend's beat-up minivan to move your new couch: not very clean, roomy, or dependable.

Tumor necrosis factor (TNF): It's impossible for your immune cells to be everywhere at once; your body is simply too big and your cells too small. To counteract this, your immune system uses "intruder alert" molecules that are part of a family of molecules called cytokines. When individual immune cells run across a virus, bacteria, cancer cell, or other bad actor, one of the cytokines they release to alert the entire immune system to the presence of the threat is TNF. Unfortunately, acute release of TNF can lead to septic shock, significant organ damage, and, in some cases, death. Further, inappropriate signaling by TNF has been associated with a host of autoimmune disorders such as psoriasis, rheumatoid arthritis, and ankylosing spondylitis. Drugs that combat these diseases (e.g., etanercept) are specifically designed to block TNF activity. The double edge of the sword of TNF is what makes it interesting to biologists, either when TNF is destroying cancer or facilitating autoimmunity.

Ultimately, the TNF work was more informative than medically useful, and the lab turned its attention elsewhere. By that time, Hwu was skilled at putting new genes into T cells and that skill was invaluable for getting around the next turn in the CAR race, work done in collaboration with the Henry Ford of CARs, Zelig Eshhar (Chapter 14).

"Zelig came to work with us on this CAR project. He had done the first CAR with a model antigen called TNP," explains Hwu. The aim of that work was to first see if a genetically engineered CAR expressed by a T cell could in fact recognize its chosen target, in this case, trinitrophenyl (TNP). That experiment involved sticking the TNP gene in a normal mammalian cell—a cell that has no endogenous fungal peptides—to determine if the CAR could see it. Indeed, it could. Such experiments are called "proof of principle," and the principle this experiment demonstrated was that CARs could be steered. The next step was to steer it to a therapeutic destination, like a tumor.

"So I started working with Zelig's group because they were having trouble getting genes into primary T cells," which was not surprising to Hwu. Dr. Eshhar was using Jurkat cells, an experimental cell line of immortalized T cells, which by nature do not stably express genes that have been taken up by transduction. "That actually was quite challenging, but since I was good at that from my TNF project, I started working with him to put the CAR into T cells."

Normal cells quit reproducing themselves after a certain number of cell divisions. Immortalized cell lines have been scientifically tweaked to divide forever. There are cells used in research labs today obtained from people who died years ago; HeLa cells are the most famous of these cell lines. HeLa stands for Henrietta Lacks, a cervical cancer patient who died in 1951, and whose cells, numbering in the trillions, are still alive today, contributing to experiments all over the world.

Hwu and his co-investigators genetically built CARs for three different targets: one each for breast, colon, and ovarian cancer. They then put these targeting systems into TILs known to target melanoma: in other words, a population of T cells that the investigators already knew to be revved up. They just replaced the steering wheel.

"It was the ovarian one that worked really well," says Hwu. "Looking back, I don't know why that worked and the others didn't. We never played around a lot with the molecule, with the bridge, and with the spacer regions, and so on. But it could have been affinity. It's not clear. But it was clear that the melanoma TILs were redirected to lyse ovarian cancer cells and that was a real 'aha!'"

It was the sweet fruit of scientific collaboration. "We made a great team back then," says Hwu. "Zelig's a really smart guy ... he brought this whole CAR concept forward." This, at a time when competition in the field was already heating up. "There were a lot of groups doing it. It was very competitive, and I think that if I hadn't helped out like that ... " But he did, and Hwu's reward came in the form of first authorship of a seminal paper in the CAR field (Hwu P, et al., *J Exp Med* 178: 361 [1993]). That 1993 paper was a milestone for CARs in cancer immunotherapy. Hwu and colleagues had instructed a cell, a living T cell, where to go and what to kill when it got there, and the T cell did just that. It was a remarkable achievement spawned by a great idea: leveraging the immune system.

"It's actually a very attractive thing," says Hwu. "The patients love it, they love the concept because they see chemotherapy as toxins and they see using the immune system as something natural that they can use to help fight their cancers." Immunotherapy represents an opportunity to contribute, an approach that can upend the prevailing feeling that many cancer patients have, which is a loss of control, of being merely a helpless, pitiable object that's being acted on.

Kudos and Crickets

There's still a great deal of work to be done. Although proven safe and effective, the ovarian CAR-T cells didn't really help any patients, one reason being that the cells seemed to get lost on their way to the tumor. "That's why we switched to other approaches, like the chemokine receptor method," says Hwu.

It turns out that arming a T cell with a CAR of choice is not enough. For the targeting to be optimally effective, the T cell needs to be relatively close to the target. This proximity in otherwise healthy people is achieved through the release of chemokines (from the Greek, *kinese*: motion), which act as breadcrumbs for a T cell to follow to a general destination. These signals are issued by a number of cell types when the body's equilibrium has been disrupted in some way—an infection, a splinter, a tumor.

The problem was that Hwu's cells seemed to be unable to see the breadcrumbs. "If you have highly trained soldiers but you drop them onto the wrong battlefield, it's not going to work at all." Therefore, Hwu set out to genetically install "breadcrumb readers" (i.e., chemokine receptors) in his T cells. As of this writing, this work is playing out. "We have several patients in the clinic we've treated with this approach. If this works, it could be applicable widely to TIL therapy, CAR therapy, [and] TCR-transduced cell therapy."

Forward-looking statements aside, at the time of Hwu's initial paper, the promise of CARs, and immunotherapy in general, continued to under-whelm the wider world of clinical medicine. Funding was scant, and the opportunities to present results, disseminate findings, and engage skeptics were few. "At meetings we were always the smallest sessions," recalls Hwu. He remembers one such session where the assigned room was little more than a classroom. "Drew Pardoll [Chapter 8] was there. It was me, Drew, Dave Carbone [of Ohio State Medical Center], a few others, giving this session in immunology in this little classroom." Sadly, the minimal space was all they needed. The same few speakers manned such sessions, and the audience comprised the same few listeners. Poster sessions for immu-notherapy were consigned to the darkest, dankest, smallest corners of the meeting halls.

"We were dismissed by a lot of people, not just biotech, but a lot of che-motherapists, which was the majority of the oncology world. They were, like, 'Are you still working on that? Really? Why don't you just inject dirt?' We were literally told stuff like that." All these years later, Hwu still shakes his head, trying to accommodate the vision: "They ... they just didn't see what we saw."

> *"They just didn't see what we saw."*

What they saw were patients alive when they should not have been. What they saw were durable responses in patients. Not everybody, not by a long shot, but some, and some was more than what anyone else had to offer. "Think about what a patient wants when they come to see you in clinic," says Hwu, "They're not after a three-month increase in median sur-vival. No patient has ever asked me, 'Can you please increase my median survival by three months?'" Yet, for the longest time that three extra months of life (as compared to the current standard of care) was all you needed to get a drug approved. "That's what would get you the great big ASCO [Amer-ican Society of Clinical Oncology] presentation: the randomized study where you did just a little bit better than the standard, but where everybody still died." Hwu, on the other hand, was seeing patients that were living extra years, not months. "It gave us enough confidence to know we were onto something."

Dark Night, and the Light

Knowing you're right can be lonely. Trying to prove you're right can be worse. "Look, being a researcher is a very manic-depressive experience in general," admits Hwu, even though he has the consistent demeanor of someone who has just returned from a tiring, but very nice vacation. "Any experiment either works or it flops, or worse, it works the first time around and then it flops."

Hwu became acutely aware of this when he couldn't get a T cell to accept the new genetic instructions for a CAR. "We'd done it in TIL cells but I was having a hard time getting the titer [concentration] up and just getting the genes into T cells." Even with the best vectors, expression of the CAR couldn't even be detected because the levels were so low.

"I remember getting totally frustrated and depressed, like, this is not going to work." But the morale of the lab kept him tethered to the bench. "We just keep encouraging each other. I remember Steve psyching me up once: 'I know you can do this. It's the kind of guy you are. You're going to get this done.' So I'm like, 'Yes, I can do this.'" The memory is palpable, and he takes the lesson home: "Part of leadership is believing in people even before they believe in themselves." Help them see what you see. Lend an ear. Take a hand. No one can do this alone.

Join the Band

The three takeaways regarding the life and times of Patrick Hwu are these: He's from God-knows-where in West Virginia; he's a gifted scientist; he's in a band. He's in The Checkpoints, a rock and blues band, where every member is a prominent immunotherapy researcher.

"I was at an ASCO once and Tom Gajewski [Chapter 24] and I were talking and we realized that we had almost identical Les Paul guitars. We both loved to play music so we decided, hey, let's get everyone together." Hwu sent out a call for colleagues to show up at the next science conference with whatever instrument they played. And they did. Someone even brought a trombone. "It was fun, but it was kind of a disaster in terms of how it sounded, so Tom and I thought, well, let's get a little more organized about this. So, I knew Jim [Allison, Chapter 1] played harmonica—we'd played together a lot when he came to the NIH for retreats and stuff—so we got Jim. Then we were on an escalator at ASCO and Rachel Humphrey [formerly of BMS, now the chief medical officer at CytomX Therapeutics], who I'd known for a long time and had done a residency with, said, 'I can sing.'

And I'm like, 'You can sing? I had no idea she could sing.' So we said, all right, let's get together. Our first practice was in Tara Withington's office in Milwaukee, and it just took off from there."

And "took off" it did, from a first practice in Tara Withington's office at the Society for Immunotherapy, Milwaukee, Wisconsin, to their most recent sold-out concert at the House of Blues, Chicago.

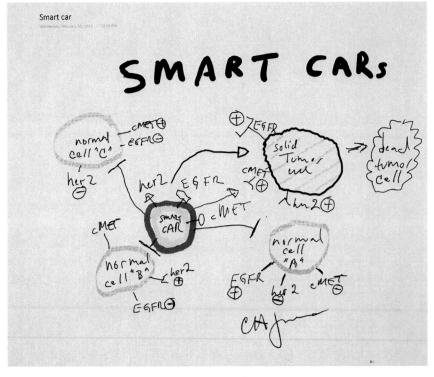

CARL JUNE
"Smart CARs"

CHAPTER SIXTEEN

Capt. Carl June, M.D. (Ret.)

Director, Translational Research Program
Abramson Cancer Center, Perelman School of Medicine
University of Pennsylvania
Philadelphia, Pennsylvania

*"The lesson I learned was, T cells could kill you more rapidly
than leukemia."* —C. June

Carl June was born in 1953 in Golden, Colorado, an idyllic setting at the base of the Front Range of the majestic Rocky Mountains. It was a great place for making strong children (June) and weak beer (Coors).

June came to science early. His father trained to be a chemist in Golden at the Colorado School of Mines, and then, as part of his army service during the Korean War, he did research on mustard gas compounds at the Aberdeen Proving Grounds in Maryland. June remembers: "He had a scar on his arm from where he had nitrogen mustard applied to show what it could do."

Note: What nitrogen mustard gas can do is cause a leisurely paced, excruciatingly painful, horrific death. It was used extensively by both sides in World War I. Ironically, as was discovered years later, nitrogen mustard compounds can also effectively treat lymphoma. These drugs are still in use today. Swords are now plowshares.

After the war, the June family moved to Emeryville, California, where June Senior worked as a chemical engineer at Shell Development, a research center operated by Shell Oil Company. Ironically, that same facility was purchased years later by Cetus Corporation, a biotech company that spearheaded the development of IL-2, a drug championed by Dr. Steve Rosenberg at the National Institutes of Health (NIH), which arguably set the whole cancer immunotherapy field in motion (see Chapter 13).

"So, that's where I grew up, with my dad working for Shell Development doing chemical engineering. Then in 1971 I got my admission letter to Stanford, because I was going to go be an engineer like everyone else in my family." A nice idea—going into the family business—but it didn't happen. June's timing was off, because the Vietnam War was still very much on. "That was during the period where your birth number was put in a lottery and depending where you came out you either got drafted to Vietnam or not," explains June. "And my draft number was 50."

Perversely, these annual war lotteries were televised, like a game show, and the viewership among the coveted 18- to 24-year-old demographic was understandably high. The lottery number assigned to June in 1971 was relatively low. It meant he was going to war.

"That year about the first 150 [numbers] went unless you either went to Canada or you were not physically qualified. And I was physically qualified." (To this day, June competes in ultramarathons.) Rather than wait for the knock at the door, June decided to enroll in the Naval Academy. "The thought was if I had to go into the military at least I was not going to be in the rice paddies as an enlisted man, but as an officer."

Choosing the Navy also positioned him (he hoped) to pursue his interest in engineering. "The best thing within the military for me was if you could be on a nuclear submarine or something. I actually spent a summer on one, a nuclear submarine out of Pearl Harbor," and he has the scar to prove it. "I was 18 years old," says June. He was eager; he was a go-getter. He was a lanky young man, well over six feet tall, with a habit of running between duties, which is problematic on a sub. "I was running and I coldcocked myself, banged my head right on top there." Fortunately, it was a short underwater tour and there was no further abuse of June's noggin.

Thereafter, course work at the Naval Academy resulted in a B.A. in the Life Sciences, followed by medical training at the Baylor College of Medicine in Houston, all paid for by the Navy in exchange for extended service. "When I got done with medical school I owed 12 years back to the Navy," notes June. This meant that his doctoring duties would focus on whatever the Navy thought prudent. This initially led to a stint in the labs of the World Health Organization in Switzerland where he worked—and then published findings on malaria, as well as HIV.

The experience brought clarity of habit. "I found out I really had a bug in me to do research." At the same time, June found out that he really enjoyed taking care of patients. These twin realizations motivated a return to Baylor for advanced training in both infectious diseases and oncology.

"I had another unexpected benefit of geopolitics," June explains, "which was the Navy had a big political problem." The Navy needed their subs to be near the enemy, and that meant deployment off the Soviet coast in places like the Sea of Japan. But having nuclear weapons parked just off their coast led the Japanese to pressure the U.S. Navy to develop contingency plans for treating victims of accidental radiation exposure.

"So they sent me and three other physicians to the Hutch [the Fred Hutchinson Cancer Research Center in Seattle] to learn bone marrow transplantation," the only effective treatment for advanced radiation poisoning. One of the other two doctors sent by the Navy was Craig Thompson, the current president of Memorial Sloan Kettering. The government sent the trio to learn the techniques and infrastructure necessary to set up a transplant center at the naval hospital in Bethesda, Maryland.

However, once again, politics intervened. "By 1989, the Berlin Wall comes down and the Navy no longer had any nuclear fears," says June. They pared down the fleet from 1200 ships to around 500, and the bone marrow transplant unit never opened. In fact, the specialized rooms that had been set up for transplant patients—highly advanced, completely sterile laminar airflow rooms—were repurposed to house Navy chaplains.

Grand Master T (Cell)

As for the stint in Seattle, the knowledge gained was still highly useful, and the experience could not possibly be forgotten. "I saw the first patients die of GvHD [graft-versus-host disease]," says June, "That was awful." At the time, there was even a protocol to increase the advantage of the immune effect, whereby patients received even more T cells than they would have prior. "And we had 10 patients in a row basically die of GvHD, so the lesson I learned was T cells could kill you more rapidly than leukemia."

The experience was not all grim, and some of the findings proved invaluable. "I had one lucky experiment," a luck facilitated by the then just-discovered technology of hybridomas, which uses living cells to manufacture antibodies. "The laboratory I was in wanted me to just make more antibodies, which was kind of like stamp collecting … Like, how many antibodies can you make?" It was high science to be sure, but the technology had quickly become a party trick—anybody could do it. The meaty part was in using these antibody tools to pry out information on biological function: in this case, the function of T cells.

June's lucky experiment involved antibodies, T cells, and a drug called cyclosporine. Approved for use in 1983, "Cyclosporine was supposed to be this miraculous cure for GvHD, because it didn't immunosuppress your whole body as much as other treatments, and it didn't kill the T cells," he explains, which is good because that would destroy the transplant. Cyclosporine didn't kill cells; it just made the T cells stop attacking you. "But when we [used cyclosporine] with T cells fully activated in the presence of one of the antibodies that we called 9.3, they didn't care," says June. "They just grew normally. And it turned out that that antibody we had [9.3] targeted CD28."

In other words, although cyclosporine told the T cells to retreat, signaling via the CD28 receptor ordering the T cells to attack proved a more potent command. June had inadvertently found the molecular counterpunch to James Allison's discovery, CTLA-4 (Chapter 1): Allison found a "Stop" button; June and his colleagues found a "Go."

"That's why from then on I studied T cells," says June, churning out more than a hundred papers since that time, many coauthored with his once Navy colleague and now friend, Craig Thompson, on the biology of CD28.

Cancer? No. First, HIV

Back in Bethesda, June's attention turned toward the study of HIV, not because cancer was no longer interesting, but because it was no longer being funded. "Where I could get funding was HIV or malaria." Why? Because of the legal separation of federal funding mechanisms. Cancer cash was bestowed by the Department of Health and Human Services directly to the National Cancer Institute, whereas, literally across the street at the Bethesda naval facility, that funding flowed from the Department of Defense, and the DOD's dictated research focus was developing treatments for combat casualties or infectious disease. June chose to work on the latter, taking immediate advantage of the enervating ability of the CD28 antibody, working quickly to establish it as the basis for a live-cell culturing system that could propagate T cells by the boatload.

The wealth of study material provided by this innovation resulted in new HIV-related information that suggested a potential therapeutic: anti-CD28 plus the T cells themselves. "We had two papers in *Science* in the 1990s about growing T cells from patients with HIV, and we found some really interesting things about the CD28 pathway." For instance, when T cells were exposed to CD28 antibodies that had been physically tethered to the bottom of the Petri dish, the cells became resistant to HIV infection. Conversely, adding a few drops of soluble CD28 to float around in a slurry of T cells made them highly susceptible to infection. This phenomenon is still being explored in the June lab.

The other therapeutic avenue that June explored was directing CD8 T cells (i.e., the killers) to attack the CD4 T cells (the so-called "helper" T cells that provide various types of support for the killers), these latter cells being the host target for the HIV. For this work, June used a CAR design. "The first CAR trial was actually in HIV patients, not cancer. A lot of people don't know that."

"The first CAR trial was actually in HIV patients, not cancer. A lot of people don't know that."

In fact, there were three clinical trials using this approach performed in collaboration with a biotech company called Cell Genesys. The results were compelling: The safety of the treatment was demonstrated, and HIV patients showed improved immune function. However, the effects were modest and before the platform could be optimized, Cell Genesys pulled the plug. "The trials were stopped in 1997 when protease inhibitors for HIV came out," says June. "So, no one cared anymore. HIV was killing everyone when we started our CAR study in HIV, but after that it became kind of like treating hypertension: if you take your meds, you're pretty much fine."

This turn of events left June holding the bag, or more properly, vials: vials full of blood samples from all the HIV patients in the clinical trials. "A major benefit that came out of those studies was that the NIH required, and still requires, that if you're doing a gene-transfer protocol [CARs are the product of genetic transfer] you have to report 15 years of long-term follow-up on the patients," says June. The purpose of this lasting scrutiny is to see if any of this *whizbang* genetic technology caused unforeseen genetic mutations.

There were no aberrant mutations in June's samples, but weirdly—even all those years after he experimentally parked them there—June could still find his CARs. "We cracked the cells, analyzed all of the patients and in fact, because we were hoping then we could just put this thing to rest and not have to do any more, it turned out everyone was still engrafted with the CAR cells." Further analysis indicated that the half-life of these cells was 17 years. June was amazed. "When Patrick Hwu [see Chapter 15] did the first CAR cancer trial at the NIH with Michael Kershaw and Steve Rosenberg [Chapter 13], with a CAR-T targeting the folate receptor α 1, those cells lasted less than a week ... I mean, drastically different."

The contrasting fates of these cells suggested that the method used to activate and multiply the cells mattered, and that the therapeutic milieu— HIV versus cancer—also mattered, greatly. "It tells you that cancer is much more immunosuppressive than HIV." In aggregate, the HIV/cancer CAR investigations told the research community, not to mention Wall Street, that CARs couldn't get you very far. What funding there was dried up.

"A few of us continued," says June. "Most people quit." Neither he, nor his CAR-innovating colleague Michel Sadelain (Chapter 17), the two foremost experts in the design and clinical use of CAR technology, could squeeze another dime out of the NCI. "It was philanthropy that rescued us." A well-heeled Penn graduate ponied up enough money to keep some CAR trials running, but funding continued to be a battle.

June could have walked away. He could have worked with more established technologies. But in 1996, it got personal. In 1996 June's wife, Cynthia, was diagnosed with ovarian cancer.

"I tried to put together a combination immunotherapy for my wife, and I actually had my own lab make a GVAX [vaccine; Chapter 8] for my wife with her own cancer cells, and we vaccinated her and I'm pretty sure it worked," says June, but it wasn't enough. Cynthia's cancer was suppressing her immune response to the vaccine. She needed a checkpoint inhibitor.

June was aware of the anti-CTLA-4 drug, ipilimumab, being developed at the time by Medarex, and he tried to obtain some under the auspices of compassionate use, a program whereby desperately ill patients gain access to investigational, unapproved drugs. "I tried very hard and I couldn't get ... It was just impossible." The FDA would eventually approve ipilimumab in 2011. Cynthia June died in 2001.

"After what happened to my wife, I started working on that full-time." "That" being the clinical translation of CAR technology to cure cancer.

CD19 CAR: The Poster Child

There are reasons that CAR technology is highlighted so extensively in this telling of the IO story, not the least of which is that using a living, genetically engineered human T cell as a drug, like some kind of self-actuated pill, is just plain mind-blowing. Also:

- The CD19-targeting CAR works beyond anyone's expectations.

- One of the first patients successfully treated with a CD19 CAR is a charming little girl named Emily. She almost died. The CD19 CAR saved her. Headlines ensued.

Choosing CD19 as a target for a CAR construct was a cross between a throw of the dice and a safe bet. CARs had shown little efficacy in the setting of cancer (the dice throw), but if it did work—and in so doing, wipe out all the B cells (cancerous and otherwise) in your body—it was known that one could survive without B cells (the safe bet).

The concern had to do with so-called "on-target/off-tumor" effects. June explains: "There was a CAR trial done by Cor Lamers and his group in the Netherlands that targeted carbonic anhydrase IX (CAIX), which is a bicarbonate pump that's overexpressed in kidney cancer." Other tissue types express this target, but not at the extremely high levels seen in cancer. It was reasonably assumed, based on prior experience using therapeutic monoclonal antibodies, that the low level of expression of off-tumor targets would not be an issue. "In the antibody field, where there's a huge amount of expertise by the Pharma industry with all these different recombinant monoclonal antibodies, you need something like 100,000–1,000,000 targets on the surface of the cell before they figure that's a good target," says June, "And so when you have 1000 or less, they don't even look."

In this case, they should have. The CAR design targeting CAIX did indeed drive right up to the kidney cancer as intended, but then it crashed into the patient's liver, causing extreme toxicity. "It was a real lesson for the field because when they went back and looked more closely, they found out there was very low-level expression of CAIX in the liver, which had been overlooked." Given this experience, the FDA stepped in and said that further CAR investigations should only be performed on extremely well-characterized targets like CD19, the thought being to iron out the kinks of the technology with a relatively safe target, and then proceed from there.

Jersey Bill

The CD19 CAR that Carl June and company spliced together was considered "relatively safe" to give to a patient that was already dying of cancer. It had a chimeric T-cell receptor targeting CD19, and a co-stimulator molecule called 4-1BB—which is very similar to CD28 in that it ramps up T-cell activity.

Although the target was well known, in 2010, the technology was not. "It was a 'fingers crossed' kind of thing," admits June. "We had no idea. We just didn't know." Previous CARs, although ineffective, were also not very toxic, so they weren't at all sure what to expect. "We knew that [the CAR] hit the target because his normal B cells also went away, and the leukemia was gone after a month. The big surprise was that it worked as well as it did."

This is not to say there were no complications. Not long after being treated, Mr. Ludwig spiked a temperature that suggested he had a bad infection, which would not be uncommon in a leukemia patient—it's how leukemia patients often die. Lab tests showed that Ludwig's body was mounting an extreme immune response against *something*, but June and the rest of the medical team could not identify the offending bug. After several weeks the situation resolved itself: The worrisome symptoms went away—as did the patient's tumor.

June teased out what happened after the fact. He did a retrospective analysis of blood samples taken during the time Ludwig spiked his fever, and the results were telling. "We got back his levels of his cytokines, particularly IL-6 and interferon γ (indicators of T-cell activity), which were sky-high in the patient and they correlated with when he got the fever, so it didn't take a genius to figure that out once we had the data, but at the time when it happened the clinicians at the bedside had never seen a syndrome like that before. I mean, it just didn't exist." However, in other CAR recipients thereafter, it happened reproducibly. "It turns out now it only happens in the patients who benefit from the treatment."

What happened is an event now called cytokine release syndrome (CRS), a condition whereby the immune system essentially tries to fight fire with fire, using high fevers and every other extreme tactic at its disposal in an over-the-top response. When Bill Ludwig spiked his temperature, June was not thinking CRS. There was no reason to, "because it didn't happen in our mice; it didn't happen in any of our preclinical studies." There were side effects that had been theorized, like a rogue vector unpredictably transforming cells, or T cells that, once they started multiplying, wouldn't stop, or a heightened risk of serious infection, "But it turned out that's not what happened," says June. Instead, CRS happened to Bill Ludwig, and even without effective intervention, he eventually recovered.

It also happened to the first pediatric patient to receive the CD19 CAR. It almost killed her.

Emily

Emily didn't die, and here's why: (1) Carl June is a consummate scientist, and (2) Carl June is a dad.

In 2010, at the age of five, Emily Whitehead was diagnosed with acute lymphoblastic leukemia (ALL). She was treated with chemotherapy but relapsed twice, at which point her parents were informed that their little girl had run out of approved treatment options.

Enter Dr. Stephan Grupp, the attending physician at Children's Hospital of Philadelphia, and Carl June. Together, as part of a clinical trial, they treated Emily with the experimental CD19 CAR. Although Emily was the first child to be treated with this new technology and a child's response to such treatment was entirely unknown, there was no choice. "If a kid with ALL has already relapsed after transplant, or if they can't get a transplant, they die," says June, and they die more rapidly than CLL patients, like Mr. Ludwig. "There's actually a higher unmet medical need in ALL."

In April of 2010, Emily's treatment with the CD19 CAR resulted in a whopping immune response: She developed severe CRS. Because of Bill Ludwig's experience, June knew what they were dealing with. However, because the condition was so new, they really didn't know what to do about it. "We initially tried high-dose corticosteroids to turn off the fever," recalls June, but that didn't work. "Then we tried TNF blockade [TNF is a cytokine]," and that didn't work either. At that point, Emily started to decline rapidly: she had multiorgan failure and was running out of time. With her parent's permission, a final directive was issued: Do not resuscitate.

"I was in Seattle giving a talk and I got an email from [pathologist] Michael Kalos, who at that point was doing our blood work analyzing serum cytokines." Kalos sent June the cytokine data on Emily. "I plotted out the data and it was unbelievable, I mean, her IL-6 levels were a thousand-fold above background." It was as if every cell in her body was on fire, fires set by the immune system trying to burn out the cancer.

"I called down to Steve Grupp [Emily's doctor] and I said 'Listen, Steve, this is what's going on.'" Emily had two cytokines at astronomically high levels: One was interferon γ, and the other was IL-6.

Now the aforementioned Dad part becomes relevant, elucidated with two bits of information: (1) June's daughter has juvenile rheumatoid arthritis, a potentially crippling autoimmune disease, and (2) in 2009, June was the president of the Clinical Immunological Society (CIS).

In 2009, Carl June, as president of CIS, gave the society's highest award to Tadamitsu Kishimoto, a Japanese researcher who developed a drug called tocilizumab for the treatment of rheumatoid arthritis (RA). The target for tocilizumab is the receptor for an inflammation-causing cytokine, IL-6. Because of his daughter's condition, June had been aware of the drug's development for some time, and he eagerly awaited its approval by the FDA for use in RA patients, an eventuality that occurred in January of 2010.

June knew there was no way to therapeutically manage interferon γ, but because of his daughter, and because of giving an award to Dr. Kishimoto just a few years prior, June know about tocilizumab—the newly approved drug to treat rheumatoid arthritis—a drug that targets IL-6. June suggested to Grupp that he try it, and Grupp immediately agreed. They gave tocilizumab to Emily and within an hour, she woke up. She was fine and, as was discovered shortly thereafter, cancer-free.

The use of the anti-IL-6 drug, tocilizumab, is now standard of care for severe CRS patients. "If you're religious, you would say it was a miracle," says June. Others would say it was a combination of things: luck, circumstance, and the perpetually prepared mind.

❧ ❧ ❧

As of this writing in August of 2017, Emily Whitehead is alive and well. If you search her name in Google images, you will find a lovely picture of Emily taken just after she had a nice chat with the President of the United States, Barack Hussein Obama. Emily had to miss school that day, so the president wrote a brief note to the teacher to explain her absence:

"Please excuse Emily from school—she was with me!"

—B. OBAMA

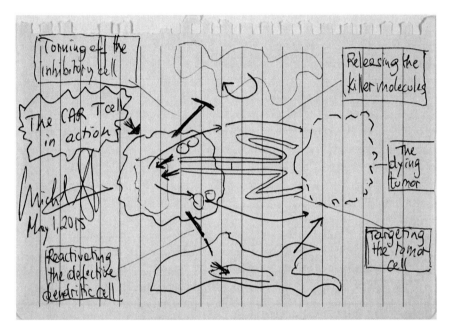

MICHEL SADELAIN
"CAR-T Cell in Action"

CHAPTER SEVENTEEN

Michel Sadelain, M.D., Ph.D.

Director, Center for Cell Engineering & Gene Transfer
and Gene Expression Laboratory
Stephen and Barbara Friedman Chair
Memorial Sloan Kettering Cancer Center
New York, New York

"My mother said, 'You know, I spoke to some of your friends, and they say you're doing this weird stuff.'" —M. SADELAIN

Michel Sadelain was born in Paris, France, in 1960, but neither of his parents were French. "My mother is from Canada and my father was a political refugee from Poland, who had been denied a visa in England, but the French gave him political refugee status." For lack of choice, and without even knowing the language, the Sadelains settled in Paris. Mom and Dad did their part and learned French, while their son expanded on their efforts by learning French, some Latin, and a lot of math, even winning an academic prize in the latter. "I was very geeky, believe me."

Not that hard to believe. Sadelain is quite tall, with a solid build, but handles himself as if he were not. His office is more thinker's nest than workspace, with journals and scientific papers piled high. The scientist sits amid them, surveying the world of ideas through glasses with thick dark frames. So, a bit geeky, yes, but not purely theoretical. For instance, Sadelain was interested in philosophy, but not the hairsplitting dialectic kind. Rather, he was drawn to the convolutions of the initiating wetware. "I really liked philosophy but I felt that it would be more informative for me personally to understand better how the brain works, and for that reason I embarked on a medical career."

If you cross philosophy with medicine, you get neuroscience, so, a direction was decided. That is, until he found something just as puzzling, just as complex. "You come across immunology in your training and you start liking it," says Sadelain, his tone suggesting that any sentient being would be similarly attracted. "I thought I would have gone on to a neuroscience career, but I then branched out to immunology." It happens. Life is rarely a straight line, usually more of a scribble.

A good scribbler makes use of the space provided, and a most excellent space was provided when Sadelain's mother's decided to return to Canada, where the University of Alberta readily accepted the Parisian prodigy.

The timing was perfect—for immunology and for hockey.

"I saw three Stanley Cups in my four-and-a-half years that I was there," asserts Sadelain, grinning, quite proud. It was a great time for hockey. "My time in Edmonton coincided with all of that, so yeah, I was pretty lucky."

Lucky in hockey, lucky in the lab; Edmonton had riches in both. In hockey, there was the Stanley Cup team, the Oilers; science had the Medical Research Council in Immunology, a collection of gifted immunologists ensconced at the university, including a number of Americans. "There was the wave of scientists who left the United States because of the Vietnam War," Sadelain explains. His Ph.D. advisor was one of them: "Tom Wegmann … brilliant guy."

All in all, it was a fertile field with good ice, and young Sadelain's talent and scientific interests grew. His skills were honed by the work he was doing with Wegmann on the mechanisms of transplant rejection, but his scientific interests were brought into single-minded focus by a lecture he heard during his graduate training by Steve Rosenberg (see Chapter 13) on something called TILs (tumor-infiltrating lymphocytes; see Hwu, Chapter 15): T cells that were being used as medicine.

The TIL work was fascinating but seemed inherently limited. Sadelain pondered a way to broaden their application, a molecular reworking of this proposed living therapeutic. "I thought, you know, we need to engineer T cells."

There is nothing like a dollop of clarity to go along with your logistical luck: Sadelain now had a singular idea. He finished his Ph.D. and immediately dove into a fellowship at Massachusetts Institute of Technology (MIT), one of the finest engineering schools in existence.

MIT

The man with the plan at MIT was Richard Mulligan, a MacArthur Foundation "Genius Grant" Fellow who, if this book were about the luminaries of genetic engineering, would be in one of the first chapters. Unfortunately, the plan he had in mind for his new fellow did not include Sadelain's overarching aspiration. Shortly after his arrival in the lab Sadelain was told, "Oh, by the way, I don't want you working on T cells, it's a really dumb idea."

Instead, Sadelain was to work on transferring genes into hematopoietic stem cells, the idea being to install genes with the instructions to make missing proteins in diseases in which there is an underlying genetic disorder. Thalassemia, a genetic disorder involving an aberrant formation of hemoglobin, is one such condition; Sadelain's project was to genetically enhance hematopoietic cells to treat thalassemia.

"So that's what I was doing there. I had an official project on thalassemia [which he works on to this day], but my underground project was really to try to figure out how to introduce genes in T cells," and this, while toiling away under a steady drizzle of negative reinforcement. "At the time, people were just learning how to insert genes in hematopoietic stem cells," explains Sadelain, and that was considered hard enough. To most people the T-cell thing seemed to be a bridge too far, so far that at one point, after a few years' worth of experiments gone awry, even Sadelain's mother chimed in. "My mother said, 'You know, I spoke to some of your friends, and they say you're doing this weird stuff.'"

Such remarks were daunting, to be sure; his staunchest supporters had doubts. "Very smart [and caring] people would just look at me, like, '*Why would you do that?*'" Hearing this, an uncertain young scientist might change his research direction. A fragile one might change careers. But a scientist who is committed to an idea might lean in.

For encouragement, Sadelain turned to Dr. David Baltimore, winner of the 1975 Nobel Prize in Physiology or Medicine, a giant not only in molecular biology but of science at large. At that time in the early 1990s, Dr. Baltimore's laboratory was at the Whitehead Institute, less than half a mile from MIT. "I spoke to him on several occasions and he didn't think it was so crazy," says Sadelain. In fact, he took an interest in the technology. At the time, HIV had just been cloned, and clinicians had finally realized that HIV was the virus that caused AIDS. "What was critically needed then was to have an animal model for AIDS," says Sadelain, but no one, Baltimore's people included, could stick a retrovirus vector in a T cell (a requirement for an AIDS model because HIV—a retrovirus—infects T cells). "People tried and they just couldn't do it."

Sadelain couldn't do it either (there is no human HIV model to this day), and the lack of progress only nourished the dispiriting opinions. "It failed," Sadelain flatly states. "It failed in one year, and in two years, and even three years later I could barely get genes into T cells."

Dark Night, Shadowed Days

How do you keep going when even your own mother has her doubts? "I think you have to … You probably have to be a *little* crazy or obsessive to go through this," says Sadelain, speaking slowly and deliberately, as is his thoughtful manner. "There's got to be that force from within that tells you that there's something here that's a greater mission, that you're going to accomplish it, no matter what. That's what I think it takes if you're going to do something different." Conversely, "You know, if you're doing things that are variations of the themes that preceded you, then maybe it's almost the opposite." In that case, craziness isn't required; you can learn what there is to learn, and organize that learning, and strategize and calculate, and successfully do all the second steps after someone else has dared to take the first. "But those people who take the first step I think have to be a little ..."

A little *off* maybe?

The speculation left unfinished, it's unclear if Sadelain considers this unnamed quality a blessing or blight. Likely both.

At the end of his travails at MIT, he eventually did figure out how to install genes in T cells, a grand accomplishment that met with near-zero fanfare. "From all of this I only had one not-so-remarkable paper," says Sadelain, and it was a thalassemia paper, no less. Such a scant publishing history should have been a red flag to future employers since publications are the coin of the realm—but somehow, word of his skills and his after-hours T-cell tinkering got out. "When I left MIT I applied to six places and I got six job offers." He settled on the offer from New York City. "When I started at Sloan Kettering that's the first thing I did: I hired a technician with my money and we started working on T cells." That was 1994, and that should have marked the beginning of a victory lap. It wasn't. It was as if all the pushback had marked him in the passing, and the career support he craved seemed sabotaged by it.

A good chunk of the problem had to do with the lack of a champion, someone to both promote and to defend this fledgling investigator. For Sadelain, this deficit was, in a number of ways, circumstantial. Shortly after finishing his Ph.D., his mentor, Tom Wegmann, died; his fellowship mentor, Mulligan, thought his ideas lacked merit; Baltimore was a fan, but played no role in his training; and Sadelain's colleagues from medical school were all physicians, not scientists (and they were in France). This lack of lineage and weak overall support network has dogged Sadelain throughout his career.

"I did my career as an orphan, and it shows," admits Sadelain. And it's cost him. "If you want to know why I'm not getting the awards [and subsequent grant support] that some are getting, it's because you need a champion, a protector, because the field was so maligned. And if you don't have strong politics behind you, it's very challenging to arrive at a certain place. This is a predicament."

It doesn't help that Sadelain sees that "certain place" as having some of the unsavory elements of a Star Chamber. Consider peer review, which is supposed to be an unbiased review of your work by your *peers* to determine if it is worthy of publication. Such validation is important to scientists at all levels of achievement: thus the phrase, "publish or perish." "I've taken a bit of a cynical view from my experience of peer review—that's what it's called—but the real meaning of it is 'the power of retaliation,'" says Sadelain, with a mirthless smile. "That is, if you don't let mine go, you know that I can get you back next time."

There's a level of science—and we're talking about that level now—where success depends on three things.

- What you did. Inspired insight is essential, but others actually need to understand what you did. Good scientists must also be good salesmen, and to be a good science salesman, you need to be a good storyteller: "Once upon a time there were some T cells needing some help …"

- Who you are. Training in a well-known lab is akin to being in a (albeit relatively small) powerful extended royal family.

- Who's behind you. As on the reality show, *Survivor*, if you want to stay on the island, you need to build alliances. You need to play well with others, or you will be punished.

Indeed, high-level, high-dollar science is not always happy folks in lab coats sitting around singing "Kumbaya." "There is a dark side to all this," says Sadelain. It's there: a high-stakes game among Olympian egos, intellects, and entrenchments. At this level of achievement, it has to be expected. Note: New York–based cancer luminaries Lucio Luzzatto and (later) Harold Varmus and Thomas Kelly would eventually provide shelter/support for Sadelain's (unconventional/unorthodox/iconoclastic) research. No small feat, that.

These days, for Sadelain (and by extension, for many others in the CAR-T space) much of the opinionated dark has given way to a great deal of warming, validating light. Consider this closing quote from an abstract regarding the engineering of T cells, presented at the World Congress of Immunology in 1992 by a young Michel Sadelain: "We hope that such genetic alteration of primary lymphocytes, conferring new specificities, regulated responses, or cytokine profiles, will provide the means of controlling immunity in an experimental or therapeutic setting."

In 2013, 21 years after that presentation to the world of his dumb idea, Michel Sadelain, Renier Brentjens, Phil Greenberg (see Chapter 12), Michael Jensen, Stan Riddell, and Isabelle Rivière (Sadelain's frequent co-investigator and wife) founded a company called Juno Therapeutics. At the core of Juno's business plan sits Sadelain's engineered T cells. The current market valuation for Juno is in the neighborhood of $2 billion.

CD28, CD19

There is a race between any number of CAR companies (Novartis [see June, Chapter 16], Juno, Kite Pharma, etc.), and it's likely, if not certain, that the first CAR-T therapy will be approved before this book sees print. Whoever wins that race to commercialization will owe their success to a number of critical innovations: the discovery of the T-cell receptor (TCR) (see Tak Mak, Chapter 11), the ability to genetically engineer a T cell (Sadelain); the idea to combine the parts of the TCR into a single genetic construct, thereby enabling their installation and targeting (see Eshhar, Chapter 14); giving the CAR a gas pedal (Sadelain); and giving the CAR a readily achievable destination (Sadelain).

"To me, the big leaps of the CD19 CAR story are the last two things," explains Sadelain. "One, the idea that you would have to create a T-cell receptor that combines multiple functions, not merely mimic a normal receptor, which is what Zelig had been doing, but something else." That something else was to take the immune-stimulating molecule that Carl June and company had investigated, CD28, and incorporate that in a solid-state fashion as one single gene into a CAR.

The first-generation CARs produced by Eshhar and others failed. They killed cancer cells, "But the problem is that killing is not enough to do in medicine," says Sadelain. "You need a T cell that not only kills but can divide and secrete appropriate cytokines and can persist long enough to actually eliminate a cancer." Simply put, first-generation CARs lacked drive. They would kill once or twice, get tired, and then stop. This condition is called "anergy." As such, "It would be a useless drug because having a T cell that functions for a few hours or a few days, knocking off a few tumor cells, that's not going to buy you an effective therapy," says Sadelain.

When Sadelain added the immune-stimulatory molecule CD28 as a gas pedal to the TCR construct, the slump toward anergy was overcome. (The CAR-T cells used by Carl June use a different "gas pedal" called 4-1BB, developed by Campana et al. at St. Jude Children's Research Hospital.) The resulting second-generation, serial killer CAR is, as Sadelain puts it, "the most complex drug ever assembled."

> *"[It's] the most complex drug ever assembled."*

The second leap was to find an appropriate target, that being CD19. (See June, Chapter 16.) "If you think back on this story now, what made the big impression on scientists, patients, investors, and Big Pharma were the obviously good (in fact, unprecedented) clinical results," says Sadelain, "And this is owed to the design of the more complex type of [T-cell receptor] fusion, and having a really good target." Had a less fortunate target been chosen, it's possible that T-cell science would have culminated in little more than some intriguing journal articles. "But CD19 was spectacular. And that's the second contribution that we made. That's what made this whole thing work."

Everybody Gets a CAR! (Actually, No)

Pharmaceutical smash hits that they are, CAR-T cells are in their infancy, and their application in the short term is limited. To date, CAR-Ts can only reliably get you to small handful of targeted places (i.e., CD19), and that limits clinical use to certain hematologic cancers (e.g., leukemias, lymphomas, myelomas). They are also difficult to make, requiring facilities that are located at only a handful of cancer research centers, and the product is autologous in nature, meaning it can only be used on the patient from whom the T cells came originally. If one used the same product on a different patient, that patient likely would die from graft-versus-host disease, or the CAR-Ts would all be destroyed by the patient's immune system.

To the first point, Sadelain, and many others, are confidently working on tailoring CAR-T cells to attack solid tumors (of which there are many more types than hematologic cancers). As to the limited dissemination of CAR-T manufacturing technologies, this is primarily an engineering challenge to be overcome by refining the existing T-cell-growing technology. The fruits of this work will likely be available within several years. Solving the third issue requires a leap of faith that one can eventually learn how to make allogeneic (i.e., genetically distinct, but from the same species) T cells, as opposed to the autologous T cells created from the cells of the patient in which they are used. This would entail extracting T cells from a healthy donor, tricking them out with CARs, and then safely administering them to an unrelated cancer patient. That utterly foreign cells can be modified in such a way that the recipient's immune system would tolerate them without using immunosuppressive drugs (which would defeat the purpose of the transplant) is a hard argument to make, and relatively few are trying to make it. Yet, Cellectis, a company in France, is using genetic manipulations to make CAR-Ts that can be used in any patient. This so-called "off-the-shelf" approach (as opposed to the current "made-to-order" paradigm) is currently being tested in the clinic, and initial results, although scant, are promising.

Sadelain, however, had his doubts. The debate right now regarding the utility of allogeneic CAR-Ts hinges on two things: (1) to what extent these cells can be made temporarily compatible with the patient's immune system; and (2) how long they need to circulate in a patient's body to eradicate a cancer.

Rather than waiting to find out, Sadelain is taking a much bigger leap of faith. He's working on making T cells from scratch so that he can control every aspect of how they are perceived by a recipient's immune system (friend or foe) and how they respond to a tumor. This, he hopes to accomplish using iPSCs: induced pluripotent stem cells.

Stem cells. Stem cells are the progenitor cells from which all mature, differentiated cell types arise. There are two general types of stem cells. First (literally) are the pluripotent embryonic stem cells, the source from which all subsequent cell types — cardiac, muscle, brain, etc. — are derived in a developing fetus. Flowing from that primordial population is the second cell type, the tissue-specific adult stem cells, which are the wellspring from which all worn-out or damaged cells are replenished throughout your life.

Induced pluripotent stem cells (iPSCs) are adult stem cells that have been exper-imentally dematured: convinced by the introduction of four crucial genes to behave like embryonic stem cells. Theoretically, iPSCs can be programmed to differentiate into any cell type (such as T cells) and then produced in mass quantities.

By using iPSCs—rather than teaching an old dog a new trick, as Cellec-tis is doing—Sadelain thinks it will be easier and more efficacious to train a puppy. "This goes back to our earlier conversation about the evolution of T-cell therapies from natural therapies [plain T cells] to synthetic therapies [CAR-Ts]," Sadelain explains. "If we drive these cells in vitro and take advantage of a genetic selection of the donors, coupled to further genetic engineering of those cells, maybe we could make cells that can be adminis-tered to many recipients." This makes both good clinical and good business sense.

"To do this we have to overcome the two problems that I mentioned before, explaining why the field is presently in the autologous mode." The techniques to design cells that elude rejection by the recipient have recently been greatly enhanced by the discovery of CRISPR-Cas9, a molecule of bac-terial origin that has revolutionized genetic engineering. As to the question of how long CAR-Ts need to remain in circulation: "I don't think any of us knows exactly what that timeline is, and it's very possible that the timeline may be different for different cancers," says Sadelain. "It may also be different in different people because of their genetic background, so maybe there's not one simple answer to all of that, but there's a number somewhere that says, 'You know, if the [CAR-T construct] can fend off rejection for so much time, that's the time window that I, the physician, have to intervene.' This is why I believe in making more potent cells that can accomplish more in a shorter time span."

The allogeneic approach using iPSCs is not all blue-sky pondering; Sadelain has already laid the scientific groundwork. "But again, we are here in the realm of the uncertain and the unknown ... what you could also call 'high risk' since the answer is not yet obvious," cautions Sadelain, who then sums up, "None of this is easy. It's a very ambitious undertaking, but that's why I'm optimistic about it."

> *"None of this is easy. It's a very ambitious undertaking, but that's why I'm optimistic about it."*

Isabelle Rivière, Renier Brentjens, Tom Wegmann

Precedence builds confidence. Sadalain can be optimistic about winning with iPS cells because, in a way, he's done it all before; the extraordinary challenges of CAR-T cells, ranging from the technical to the clinical, were eventually overcome—overcome with an obsession for the idea, and the help of three gifted investigators: the first two in the present, and the last in the past.

Isabelle Rivière: Sadelain met Dr. Rivière in the early 1990s at MIT, where her work led to the creation of one of the most utilized retroviral vectors in research and medicine. This technical know-how was invaluable to the nascent CAR-T program because, "Everything had to be built from ground zero," says Sadelain. Truly everything. A facility for T-cell production had to be designed and built; personnel with no experience in such work prior (because there was no such prior work) had to be trained; regulators had to be educated; assays, metrics, bells, whistles... .

"Dr. Rivière was amongst a handful of researchers in the world with expertise in immunology and genetic engineering ... so crucial to the success of clinical translation in the academic setting," says Sadelain. "We could not have demonstrated the efficacy of CD19 CAR therapy without her guidance."

(Full disclosure: Although Sadelain and Rivière initially shared only their intense passion for clinical research, they now also share an apartment and two kids.)

Renier Brentjens: Dr. Brentjens was a fellow in Sadelain's laboratory, where he performed the initial animal experiments testing CD19 CAR-T cells, and where, now, he likes to recall that, after he reported the first-ever data with the first-ever, still-living, genetically engineered T cell capable of curing lymphoma in mice that, well, um, ... "Nobody cared." Yet, he cared, and had the courage to lead the first-ever CD19 CAR trial at MSK, even though he had to fight to get colleagues to enroll their patients on his CAR-T trials and, further, to convince patients to take part in an underappreciated, unprecedented, nearly unfathomable experiment.

The result now being waiting lists for CAR-T treatment.

Tom Wegmann: Shortly after young Michel Sadelain received his hard-won Ph.D., his mentor and champion, Tom Wegmann, had a stroke and died. He was 53.

"I liked Tom enormously. He was such a jovial person, boundless energy, and he had this willingness to entertain wild ideas … if they were well argued, of course."

Wegmann trained at the University of Wisconsin under the Nobel laureate, Oliver Smithies. After earning his academic stripes he was courted by not one, but two, Nobel laureates: David Baltimore, and James Watson, co-discoverer of the structure of DNA, who was then at Harvard. "So these two guys are fighting over this brilliant young man," says Sadelain, "But it's also the time of Vietnam so, like a number of scientists, he left for Canada and the University of Alberta."

The venue suited him in any number of ways. For instance, Wegmann was something of an outdoorsman. "He would go ride horses with [fellow immunologist] Irv Weissman [of Stanford] who came frequently there. I think they were hunting for beavers, maybe … I'm not really sure what they were doing in the mountains all the time."

Wegmann's sense of adventure extended to his research interests in immunology. "He had these crazy projects on the role of alloreactivity in pregnancy, which to this date remains a big topic regarding the role of the immune response in promoting the growth of the fetus." (See Munn, Chapter 22.) The research took Wegmann to some interesting places. "He was going to Saudi Arabia to advise the King who had problems breeding his camels to win races because they were so inbred that they had repetitive abortions." Wegmann tried to figure out how to stop it. Wegmann also worked on malaria, trying to figure out how one might block transmission of disease by genetically altering mosquitos.

Then, there was the sewing together of mice. Sadelain pitched in on that little project. "It's called parabiosis, when you take two mice and you sew them together by the waist through the skin." The physiologic result of this surgical pairing is that the two mice then have a single circulatory system. Essentially, you create rodent Siamese twins.

Question: Why in God's name would anyone do such a thing?

Well, not surprisingly, most of the mice developed from GVHD, but, surprisingly, some did not. The surviving mice justified the procedure, which was intended to explore mechanisms of transplant tolerance: the process of getting an immune system to accept a foreign implant. "What they found out is that in the healthy mice, one of them would have completely engrafted the other in the bone marrow," explains Sadelain. "So, a spontaneous bone marrow transplant had occurred from one to the other, without any chemotherapy or radiation." The two mice now had identical immune systems.

The implications of this were profound. Wegmann worked hard to unravel the mystery. "He realized that there was an antibody response [an immunologic attack] of one mouse against the other. If the antibody response was slow, the two mice would go into a graft-versus-host mode and kill each other. But in some of these pairings, one of the animals would respond against the other, and these [responses] were antibodies against MHC [major histocompatibility complex; a very important molecule involved in recognition of 'self' by the immune system]. And that allowed for the marrow of one mouse to engraft the other."

If you could do the same trick in people, it would revolutionize the field of transplantation.

"If you could do that, figure it out, it would have a profound impact," says Sadelain, who worked on this question in Wegmann's lab. "Because right now the way transplants are done is by using chemotherapies, radiation, and so on. But what if you could do what all that does but in a very stealth way where you could almost do it as an outpatient without toxicity? That remains a dream, but you see Tom's story from 30 years ago is just making its way in biotech circles today." In fact, two companies have recently formed to expand on and possibly commercialize Wegmann's observations.

Back in the early 1990s, people said it was a wacky idea. They said the same about Sadelain. From Wegmann's obituary in the *Journal of Reproductive Immunology* (which Wegmann cofounded): "A man of large personality, Tom was a constant source of stimulating, often controversial, ideas, delivered to his listeners with boundless enthusiasm. He was equally at home expounding his views in the lecture theatre, in the laboratory, or in a bar."

❦ ❦ ❦

On the potential of CAR-T cells:

"Michel pulled me into his wonderfully cramped conference room and showed me the data that was emerging and I almost fell off my chair. My first reaction was that it was probably not true—that's the first reaction when one shows you such data. But if we could collectively move the field forward so that we can develop this technology that will enable us to target patients with solid tumors, this would be not just transformational, this would be one of the most disruptive technologies in the history of medicine."

— José Baselga, Physician-in-Chief, and Chief Medical Officer, Memorial Sloan Kettering

BUSINESS AT THE BENCH: ONE PROTEIN, ONE VIRUS

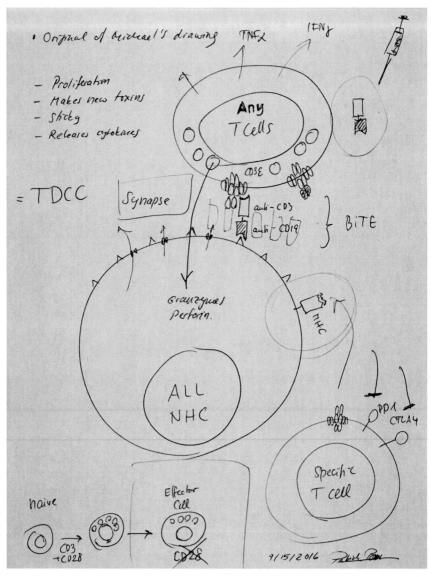

PATRICK BAEUERLE
"First BiTE"

CHAPTER EIGHTEEN

Patrick Baeuerle, Ph.D.

Managing Director
MPM Capital
Boston, Massachusetts

BISPECIFIC ANTIBODIES

"NK cells are not as good killers as T cells are. NK cells are black powder; T cells are dynamite."   —P. Baeuerle

Patrick Baeuerle was born in 1957 in Friedrichshafen, Germany, on the shores of Lake Constance, a setting of sublime picturesque beauty. "Well ... it is beautiful if you plan on retiring," says Baeuerle, trying to be fair. "But if you're a young person with a lot of great ideas ... I took the first possible chance to leave."

That chance would not come until Baeuerle was old enough to apply to university. In the meantime, he indulged his interests in the ways most young men do when growing up in such an idyllic setting: first, by developing a fondness for sketching the natural world around him—thinking that perhaps one day he would become an artist—and, second, by ardently applying himself to the art of blowing stuff up.

"Biology in the '60s and '70s wasn't really well understood," explains Baeuerle; it seemed so convoluted. "But chemistry was simple, you know, the atoms, the molecules." Chemistry was also satisfying: You could combine a pinch of this with a gram of that and squirt in some solvent and give the flask a swirl and—*presto*—something obvious happens.

"In particular, I found the exothermic reactions were fascinating." (Endothermic reactions are those that absorb heat, like the melting of an ice cube, whereas exothermic reactions release heat, as in, potentially, *kaboom!*) "Exothermic means that you mix some things together and you get a lot of sparks, and glowing, and dripping molten glass, and things like that." By the age of 14, Baeuerle had already established a lab in his parent's garage.

It was all great fun and he was good at it, and it was a great way to meet girls. "I started giving lessons to my fellow students who were not so good in chemistry, and particularly the girls were fascinated by all these reactions. They had great laughs, and everyone got really excited about things that went *poof.*" He was the neighborhood magician, a self-taught alchemist procuring his powders and potions at the local pharmacy after convincing the proprietors that he knew exactly what he was doing. Which he did, more or less.

Occasionally, less.

"One time I made an explosive you could actually ignite by adding a drop of water," says a proud Baeuerle. He made a big handful of it one afternoon and marched outside for a confirmatory field test ... just as it started to rain. And *POOF!* "Up it went and literally the skin came off my hand." Painful? Oh yeah, "but it was quite interesting."

What was most interesting to Baeuerle was not the destructive potential of the magic trick, but that it took so few ingredients to perform: sulfur, oxygen, carbon, a few other common elements. Just a handful of atoms—so much power right there in his hand—the power to destroy, the power to create, to create life itself. "Just look at carbon," says Baeuerle. The energy from the sun drives a reaction that turns carbon dioxide into a plant. A few iterations later, a plant can evolve to become a fish, a bird, a mammal, or you. "So it's all in there, all the information of becoming an important biomolecule is in the carbon atom that was created by a star, and you don't need many sorts of these atoms to come up with a human being. Isn't that fascinating?"

> "All the information of becoming an important biomolecule is in the carbon atom that was created by a star ... Isn't that fascinating?"

In fact, the remarkably short chemical inventory of a person is this: oxygen (65%), carbon (18%), hydrogen (10%), nitrogen (3%), calcium (1.5%), phosphorus (1%), and sulfur (0.3%), accounting for 98.8% of the recipe. The other 1.2% is represented by a few dashes of potassium, sodium, chlorine, copper, and magnesium.

The Getaway

Baeuerle's attraction to the art of demolition was that of a tinkerer's curiosity; he had no aspirations regarding the strategic use of exothermic reactions. When it came time to sign up for military training (as was compulsory in Germany at the time), he declared himself a conscientious objector, thereby exchanging a short stint in the military for a longer spell in civil service. In his case, that meant working for two years in a hospital.

"I was helping take people apart."

Meaning, autopsies?

"No. I just helped clean up all the stuff they took out." Regular bodies mostly, but sometimes crime victims. "So that was an interesting experience."

Having proven himself worthy in that morbid milieu (by not throwing up), Baeuerle found himself occasionally replacing the hospital photographer in surgical theaters. "I had to gown up just like a doctor and stand there, and then the surgeon would very proudly hold up some freshly removed tumor and I had to take pictures of that." Baeuerle was also introduced to the ministrations of the radiology department, as well as the department of nuclear medicine.

The sum total of his civil service experience convinced him to set his sights firmly on a career in medical science. Baeuerle obtained an undergraduate degree from the University of Konstanz (also on the coast of Lake Constance) and then finally escaped that placid pond by pursuing graduate training at the Max Planck Institute in Munich. There, he began the delicate and vital task of choosing his first mentor. "I actually walked into that institute, looked around, and then saw a big sign: Neurochemistry," recounts Baeuerle. "And I thought, that sounds interesting, so I walked straight down the hallway, turned into the first lab and met a young guy, maybe 30 years old, just back from his postdoc at Yale." His name was Wieland Huttner (currently the Director at the Max Planck Institute of Molecular Cell Biology and Genetics). "The idea just resonated well with me: neurochemistry." The two spoke briefly. "An hour later, I was his first student."

Huttner's research focused on a type of enzyme called a kinase and on the amino acid tyrosine, one of the chemical building blocks of proteins. Tyrosine kinases attach a phosphate group to a tyrosine within a protein, and when proteins are thus phosphorylated the whole cluster of atoms acts as a relay station for cell signalling. Mutated kinases often are involved in the development of cancer, an important fact recognized by the Nobel Prize committee in 1989, when the award in medicine was given to the team that discovered the proto-oncogene "src," a tyrosine kinase. Baeuerle's thesis project was to elucidate further the process by which these various molecules—kinases and tyrosines—interact.

The project took three years, three very fruitful years thanks to the right project and the right mentor. "[Huttner] taught me everything about science far better than anything you can learn at a university," says Baeuerle. "He taught me how to write papers, how to design experiments, everything." And the rewards flowed both ways: "I had eight first-author publications with him as a student, which is a rare thing."

The kinase work gained the lab enough recognition that Huttner was offered a position at the European Molecular Biology Laboratory (EMBL) in Heidelberg, a premier European research institution, which he accepted. This created a bit of tension in the lab, as everyone had to relocate 200 miles away, and

Baeuerle's wife couldn't come; her career was in Munich. Fortunately, it was a brief separation, just a year, and the time was invaluable to Baeuerle's vocation. At EMBL, he finished his thesis and he met David Baltimore, winner of the Nobel Prize in Physiology or Medicine in 1975. (A junior researcher meeting David Baltimore is like a beginning music student meeting David Bowie, or someone who just mastered their first card trick meeting Houdini.)

Baeuerle was convinced he needed to continue his training in the United States and he hoped that knowing Baltimore would be his ticket out. He applied for postdoctoral positions at five institutions and was rejected by all but one. "The only rejection I did not get was from David Baltimore. He was the only guy who wanted me. Isn't that amazing?" Baeuerle packed up his family—he had a son by now—and headed to the Whitehead Institute in Cambridge, Massachusetts.

Baeuerle immediately noted Baltimore's depth of perception. "He needs a few seconds, or maybe a minute, to understand something," says Baeuerle. "He comes to you at your bench, he asks some question or other and he immediately gets it, and he doesn't forget anything he has seen. Sometimes, if he's really excited about something he would go with you to the darkroom and wait until the film comes out of the machine. He'd grab it and say, 'Get me a slide.'"

Here We Go: IO

Baeuerle's project with Baltimore was to learn all there was to know about NF-κB (pronounced: en-ef-kappa-bee) a protein that seemed to be associated with multiple functions—particularly those of immune responses—but no one knew how the effects occurred. Before Baeuerle joined the lab, Baltimore's team had used a gel shift assay (a technique that uses radioactive DNA) to establish a tantalizing clue: NF-κB bound to highly specific genetic patches.

This association between protein and gene was not a unique observation; proteins are often used as switches to turn genes on or off or to adjust their level of production. As such, NF-κB is in a class of molecules called "transcription factors," molecules responsible for initiating the instructions as encoded in the DNA. (Think of the cell as a store where everything is made to order; you only make the product when you need it. Transcription factors place the order.) This, however, was different: NF-κB seemed to affect so many important things, so much so that the most critical initial question to ask was: What activated NF-κB?

"That was my starting point," says Baeuerle. The end point, reached two years later, was the identification of that activator. "I found that NF-κB is sitting in the cytoplasm." (Higher-order organisms have cells with two general compartments: the nucleus, where DNA is stored, and the cytoplasm, where all the machinery for making proteins according to genetic instructions is maintained.) "So NF-κB is kept there in the cytoplasm in an inactive state by I-κB [inhibitory-kappa-B], which I discovered." Only when I-κB is phosphorylated can NF-κB cross from the cytoplasm into the nucleus to tell DNA what to make. "The special thing about NF-κB is it's sitting there as an inactive precursor in the cytoplasm," explains Baeuerle. The inhibitor responds to warning signals from the immune system, like the detection of a virus, by letting NF-κB off its leash, allowing it to breach the nucleus and activate the genes needed to mount an immune response.

There and Back Again

The postdoctoral work with Baltimore attracted no small amount of attention. Baeuerle was invited to return to Max Planck to head up his own lab. "You wouldn't believe it, but three months into my own lab I submitted my first own *Cell* paper, and it got accepted!" It was read widely, and opened a floodgate. "I think I made 100 papers, and all of these NF-κB papers made me the most frequently cited German scientist of the '90s. You may have heard that."

Others certainly did. More opportunities were offered. Baeuerle accepted the most prestigious of these, the biochemistry chair at the University of Freiburg. "Boom!" says Baeuerle, "I was 34—the youngest of the whole faculty—and I had to deal with 44 colleagues running huge clinics and driving luxury cars and making a lot of money."

So, was Baeuerle really that good or just lucky: always in the right place at the right time? "Oh, there's a lot of luck behind it, plus the intuition to pick the right option when there is an opportunity. A lot of intuition, gut feelings." His approach, although, was not entirely visceral. In simplest terms, he pursued what he found most interesting. "There's not much logic behind it," or so Baeuerle says. "It's more creating the opportunity, and then pursuing it by intuition. But in hindsight, it's all a straight line. Isn't that interesting?"

Baeuerle seemed well-positioned, career-wise, when he was invited to present his data at Tularik, a biotech company based in San Francisco. "A friend of mine leased a convertible and we were driving over the bridge and the fog lifted, and there was blue sky, and these red cables, and the bay." He was enthralled. The next day he gave his lecture at Tularik. On the way out the door the CEO offered him a job. He took it. He left a highly prestigious, well-paid-for-life academic position and relocated his family to the City by the Bay. Older colleagues thought him a fool. The younger ones ached for such freedom.

The work at Tularik focused on finding drugs to discipline Baeuerle's poster child, NF-κB. They didn't find any, but Baeuerle did find an invaluable education about the industry. "I learned biotech from scratch," Baeuerle says. "A lot of it I learned from Dave Goeddel, Tularik's CEO." For both personal and professional reasons, Baeuerle left Tularik in 1998 and returned to Germany to take a position at a company called Micromet.

Micromet and BiTE

When Baeuerle joined Micromet, they only had two compounds on the table, both in-licensed—neither discovered in-house. (This is common practice: Somebody discovers something, often at a university, and then for a fee grants a license to a company for the discovery's commercial development.) Not long after joining the company, these licenses were rescinded. "In my despair, I went to the Institute of Immunology at Munich and met their people who had developed something called bispecific antibodies engaging T cells," says Baeuerle. Call it an act of desperate intuition. "Although there was a very bad history to those types of molecules, I immediately found this one very appealing." He licensed these bispecific T-cell engagers—which he dubbed BiTE—special antibodies that could usher any T cell, regardless of its prior training, to a target of Baeuerle's choosing. The innovation was to leverage a part of the T cell's guidance system: a part of the T-cell receptor (TCR)-targeting complex called CD3 that is exactly the same on all T cells (discovered by Mak, see Chapter 11).

> *"In my despair, I went to the Institute of Immunology at Munich and met their people who had developed something called bispecific antibodies engaging T cells."*

"Actually, regular antibodies are Nature's bispecific antibodies." Baeuerle explains: Picture the antibody as being Y shaped. The tips of the two prongs of the Y are designed to bind to a specific target antigen. Once bound, various types of immune cells will latch on to the stem of the Y by way of the Fc receptor (for "Fragment, crystallizable," named—apparently—to reflect the fact that the fragment can be crystallized). This allows the antibody to usher immune system cells to within proximity of the target, thereby facilitating target cell eradication.

Sometimes, however, it's not all that efficient. The immune cells most commonly recruited in this way are natural killer (NK) cells. However, as far as combatting cancer is concerned, NK cells lack the required firepower. "NK cells are not as good killers as T cells are," says Baeuerle. "NK cells are black powder; T cells are dynamite."

BiTEs (and other similar bispecific constructs) are artificial antibody-like molecules constructed to recruit T cells. It works like this: Sever the targeting

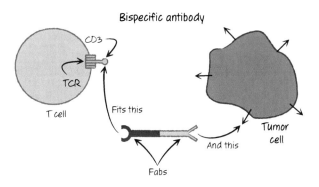

prongs from the Y of an antibody that recognizes a tumor antigen (in this case, CD19), then attach it end-to-end (i.e., stem-to-stem) to another antibody whose prongs target the CD3 molecule on T cells. The result is a short adaptor with sticky ends; one end sticks to an antigen on a tumor cell, and the other to a T cell.

Baeuerle's drug—this first-to-market bispecific antibody that targets CD3 on T cells and CD19 on cancerous B cells (thereby bringing the two together)—is called blinatumomab, and preclinical testing of the construct showed it was highly effective. After exhaustive investigations at the bench, followed by early phase clinical testing, the biotech giant, Amgen, purchased Micromet in 2012 for $1.16 billion.

The drug hadn't even been approved yet—Baeuerle and most of his team were retained for that purpose—but that's how impressive were the data. Two years later, blinatumomab was approved by the FDA for patients with B-cell acute lymphoblastic leukemia.

"That day was my wife's 60th birthday, and it was 9 o'clock in the evening and I was just about to toast her." Baeuerle can still see it. "I had a glass of champagne in my hand when my mobile rang and I saw it was the CEO of Amgen, and I thought 'What does this guy want from me at this hour? Am I fired, or what?'" So he answers: "'Hi Bob, this is Patrick. What's going on?' And he says, 'Patrick, I just got a phone call from the FDA. Your drug just got approved and you're the first one to know.'" Baeuerle could barely hold the phone and did spill the champagne. "I was shattered, I started crying and then for the rest of the evening I was just calling up colleagues to tell them what happened." This was two-and-a-half months after the drug was submitted for approval: the fastest drug approval ever by the FDA.

Slightly less than one year later, on Baeuerle's birthday, November 24, blinatumomab received its European approval. "Isn't that amazing?" wonders Baeuerle, seemingly ever cheerful. "Sometimes things just come together."

Like magic.

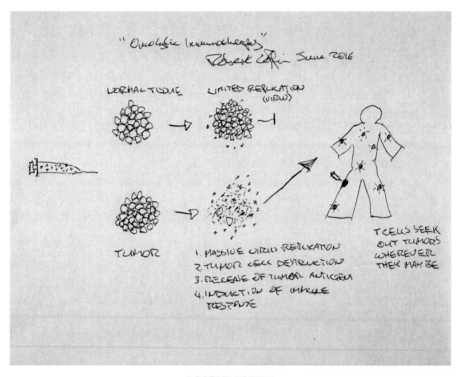

ROBERT COFFIN
"Oncolytic T-Vec"

CHAPTER NINETEEN

Robert Coffin, Ph.D.

Cofounder, CEO
Replimune
Woburn, Massachusetts

ONCOLYTICS

"It's very important to know when to walk away."   —R. Coffin

Robert Coffin was born in 1965 and raised on the south bank of the River Thames in the town of Windsor, a fairly small town with at least one very large house. "The main claim to fame is Windsor Castle," says Coffin, and for good reason. Windsor is the largest inhabited castle in the world and has served as a summer residence for the royal family for more than 900 years. "As a child, my mother used to work there in the royal library selling prints and things. I worked there too, during school holidays selling postcards to American tourists."

Windsor can be noisy, as it lies directly under the flight path to Heathrow Airport, one of the busiest airports in the world. Coffin particularly remembers the Americans for their insightful questions regarding Windsor's illustrious history, such as: "Why did they build the castle right under the flight path to Heathrow?" To this, Coffin had no polite answer.

Coffin's early education in Windsor was uneventful. "I went to Windsor Grammar School, an all-boys school." There was a Windsor girls' school as well, but it wasn't that close by, and there wasn't any formal interaction between the schools. "I didn't meet girls until I went to university, really." Freed from the demands of a social life, Coffin focused on science studies at Windsor Grammar—biology, chemistry, physics—leading to an eventual degree in microbiology from the University of East Anglia a few years later.

To this point in Coffin's tale there was no mentor, and despite what might be seen as a science focus, having an undergraduate degree in microbiology is sort of like having a degree in English literature. Books are nice in general, but if you want to write one you really need to settle on a specific topic.

"I didn't know where I was going," Coffin admits, "I was pretty direction-less until I started to do my Ph.D." Further, before he entered graduate school his grades were just so-so. "I got a 2.2, which is a second-rate degree, and gen-erally wouldn't be thought good enough to go on to do a Ph.D."

In the English grading system, there are first-class honors (1st), second-class honors, upper division (2.1), second-class, lower division (2.2), third-class honors (3rd), and "pass." The cutoff for entry to graduate school is usually 2.1.

Note: The man profiled here with the mediocre grades is the current CEO of a multimillion dollar biotech company, Replimune, and before that was the cofounder and CEO of BioVex, a biotech company that was sold to Amgen in 2011 for a billion dollars.
Just imagine if he had gotten better grades …

"I found that being taught stuff was a little boring, whereas doing stuff yourself … I became very driven after starting my Ph.D."

Graduate work commenced in 1988 at University College London, where aptitude, as well as attitude, proved to be everything. "My chap just basically let me get on with it, and I did get on with it. And that was the best thing." Coffin did not want or need overt guidance from his supervisor; he wanted the freedom to fail. "In my view, a Ph.D. should be a period of autonomous research where you sink or swim." This go/no-go decision tree is typical of Coffin, but somewhat at odds with the ultimate goal of academic institutions or departments where the drowning of graduate students is con-sidered institutionally unsustainable.

Yet, Coffin remains adamant. "Ph.D. students are often far too micro-managed, spoon-fed. They really should be just left to get on with it for three years and if they succeed, that's fine; if they don't, then they probably aren't cut out for doing basic research anyway."

Coffin was cut out for it and, left to his own devices, pursued the study of plant virology, one of the hot topics of the day. "Plant genetic engineering was coming to the fore at that time as something which was going to save the world," says Coffin "It was a new frontier and things developed very, very quickly." Innumerable projects were proposed: not just cures for diseases that affected crops, but genetic enhancements to make plants drought- or insect-resistant or to even produce human proteins in plants for use as therapeutics.

Much of what was proposed did not pan out. "For some plants, it was rela-tively easy to make transgenics. For other plants, it was much harder." And there was a bit more to it: "Unfortunately, the potential hasn't really been lived up to for improving the lives of people all over the world, improving nutritional policy and things of that sort, which it really would be great for," says Coffin, clearly disappointed, if not a touch cynical. "I think [the technology] is greatly underutilized, and for not good reasons due to fears of GMOs [genetically modified organisms]."

Coffin's Ph.D. project was cucumbers. "There's a strange virus called beet pseudo-yellows virus which causes big problems for cucurbit [cucumber, gourds, squash] production in Southern Europe and in the UK," says Coffin. "And my Ph.D. was about characterizing this particular virus."

Viruses. Viruses are really, really small. If a typical microbe were a basketball arena, a typical virus would be the ball. All viruses depend on other cells to reproduce. A virus either attaches itself to a cell surface (that of an animal, plant, or microbe) and injects its genetic material (DNA or RNA) into the cell, or the viral particle is brought into the interior of the cell intact, wherein the genetic material is released. Either way, the infected cell then involuntarily expends its own resources and molecular machinery to manufacture all the various components of a new viral particle. These components then self-assemble and are discharged from the cell in one of two ways: either in lysogenic fashion (whereby the host cell survives) or in a lytic process, like rabies, where the host cells are destroyed on release of mature virus.

The object of Coffin's investigations turned out to be a type of closterovirus, an RNA-based viral entity. Solving this minor mystery was a major boon to Coffin's skill set and a solid backgrounder for Coffin's next idea for a project: trying to use another RNA virus—the first virus of any kind ever discovered, the tobacco mosaic virus—to create a vaccine against HIV.

The idea was to vaccinate against all the strains of HIV by creating a library of tobacco mosaic virus particles containing all the genetic variations of HIV proproteins. "Because HIV is a quasi-species, there are lots of genetic variants," explains Coffin, "And that's why it easily evades the immune system and rapidly becomes resistant to [interventions]."

A __proprotein__ is a protein that does not function until part of it has been lopped off. Like Chinese takeout chopsticks, they're provided intact because they're easier to manufacture that way, but you have to break them apart for them to be useful. The enzyme that HIV uses to cleave its proproteins is called a protease; many of the drugs that treat HIV infections are protease inhibitors.

A biological library is a collection of DNA fragments that are stored in the DNA of standardized microorganisms. This allows for the ready production of genetic material for any number of applications, as in this case, the manufacture of a vaccine. In theory, one could induce the manufacture of all possible HIV component proteins from your library in a plant, harvest the plant, purify the proteins, and use the whole mélange as a vaccine.

Coffin applied for a grant to do the work. The grant was declined. The following year, he resubmitted the grant. Again, it was declined. "I think I might have got it the next year, but by then it was too late," says Coffin. "I'd already recommitted myself to working at University College London on herpes simplex virus, which is really what I've been doing ever since."

The Promise of Herpes: Parkinson's and Beyond

"Yay, herpes!" said no one—ever. However, it's worked out well for Coffin. "The project was originally to study herpes simplex virus [HSV] latency," says Coffin, "the mechanisms by which it stays dormant in an individual for a long time, but then intermittently pops out and causes cold sores." This "popping out" refers to how HSV spends part of its life cycle: the DNA resides in a highly stable circular form in the nucleus of an infected cell, side by side with the DNA of the host where, for months or years at a time, it does nothing. Then, often during times of stress, the herpes DNA becomes active and you get a cold sore.

Summing up the totality of his research into this DNA reactivation phenomenon, Coffin offers: "It's still not understood how that works." Therefore, for lack of progress, Coffin's team shifted focus to something more immediately applicable by using HSV as a vector for gene therapy. After all, if HSV can successfully deliver its own parcel of DNA into host cells, why not use it to ferry in genes of one's choosing?

"We looked at using HSV as a means of delivering genes to the nervous system for gene therapy for things like Parkinson's and other neurologic diseases." This made sense because HSV preferentially infects neurons and can remain in the neuron's cytoplasm for the lifetime of the cell. "That suggested the potential for very long-term gene expression," says Coffin, "And therefore, long-term [impact] on disease if you could exploit that in a vector." To that end, Coffin received a grant from the Parkinson's Disease Society to try using an HSV mutant, a mutant with the neurotoxic ICP34.5 gene removed, to deliver DNA to treat Parkinson's.

"It sounded like a good idea," says Coffin, "It sounded exciting to me, but it became apparent relatively quickly that while [the mutant] was nonpathogenic, it was still pretty toxic to neurons, and therefore they didn't actually survive in the long term nor give you expression of the gene you wanted in the long term."

It also became apparent to Coffin and colleagues that when you drop a foreign gene into the HSV gene pool, the ripples tend to cause the virus to enter latency after infecting the cell. "As all the HSV genes get turned off, the gene you've inserted into HSV also gets turned off." Or nearly so. "You just get a short-term burst of expression, which isn't going to treat this disease in any useful fashion."

There were two problems to overcome: How to get long-term expression of the gene of interest, and how to engineer a version of HSV that was non-toxic. "Both issues to a greater or lesser extent were resolved and we filed patents around that," which Coffin, and co-investigator David Latchman used in 1999 as foundational intellectual property (IP) for a new company: BioVex, née NeuroVax.

BioVex and HSV

The core of the new company was an optimized version of the HSV vector for DNA delivery to the nervous system. "It was things like Parkinson's disease, other neurodegenerative diseases, spinal injury repair, and chronic pain … those were the targets we were originally going after."

Although the lead program was indeed for Parkinson's, with its validated target proteins like glial cell–derived neurotrophic factor or tyrosine hydroxy-lase, Coffin thought his business would more easily move forward by working first on neuronal genetic diseases in children. "We felt that it might be a lower barrier to entry to initially try and treat children with inborn errors of metab-olism diseases which affect the nervous system and are untreatable," says Coffin, adding that these patients often die in their early teens, so there was a substantial unmet medical need.

"We were moving towards clinical trials in the nervous system with the first round of venture capital funding we raised, but about a year in we came to the conclusion that while it was very interesting and exciting, the route to getting through clinical trials was going to be very long and arduous," says Coffin. "So, we took a step back."

It was a no-go.

Note: A so-called "no-go," as used in this and similar cases, does not necessarily reflect a lack of efficacy for the proposed intervention. Rather, sometimes it reflects the fact that (a) proving efficacy is cost-prohibitive because of the number of patients required for clinical trial results to reach statistical significance (it is a business after all);

(b) proving efficacy is time-prohibitive (you really can't wait five years to see if the drug is working; it is a business after all); or (c) proving efficacy would involve measuring something that is too difficult to detect given current biological assessment technologies. These are all no-go situations that have nothing to do with the drug's actual efficacy.

Then Came T-Vec

BioVex did not have a viable therapeutic agent as yet, but they'd developed some efficient and highly versatile tools. "So, we changed the direction of the company," says Coffin, rather matter-of-factly. They went in the direction of treating cancer.

"I think it's very important to know when to walk away," Coffin says, and to do that you have to have the courage and good sense to generate definitive, go/no-go decision informing data as soon as possible. "There's a considerable tendency to not do definitive experiments, particularly in clinical trials," says Coffin. "There's a tendency to not design trials that get you a definitive answer, but allow you to wave your arms around and carry on with your gravy train."

Coffin's disdain for such investigations is clear; he doesn't like to waste time. His thumbs do not twiddle. Are you heading toward a wall? Downshift and pivot. It's a no-go? Hit the brakes. "As a small company, you can change direction quite rapidly without too much complication as long as you've got support of

> *"As a small company you can change direction quite rapidly."*

investors, which we did." The neurology programs continued on the back burner (eventually to burn out) and the focus became developing an HSV-based oncolytic, a therapeutic eventually named T-Vec.

Oncolytic. A fairly literal term, 'onco-' for tumor, and '-lytic', from the Greek lyein (to loosen, or dissolve). Thus, oncolytic therapies dissolve cancer cells. This approach has been under investigation for a number of years with a variety of virus types that naturally lyse their target cells after infection and replication.

Why would HSV preferentially target tumor cells and not healthy cells? It doesn't. Targeting occurs by default. Cancer cells, by definition, are broken, and part of what's broken is the natural defense system that normal, healthy cells use to combat viral infections. In a tit-for-tat response to these natural defenses, some viruses — HSV being one — have evolved mechanisms to overcome them. To ensure that T-Vec (Talimogene laherparepvec, Coffin's HSV-based therapy) spares healthy tissue, the evolved proviral infection strategy of herpes is engineered out. Genetically configured in this way, and taking advantage of a cancer cell's intrinsic dysfunctions, T-Vec effectively and selectively targets tumors.

Coffin was aware of the less than stellar track record for oncolytics, but he had what he thought was a better idea, a way to escalate the therapeutic response. To do that, he leveraged what happens after the cancer cells are virally "popped," an event and aftermath that suggested the potential for a vaccine. "People have been trying to develop cancer vaccines for many, many years with limited success," says Coffin, but there was one approach that did show promise, and that was autologous vaccines: vaccines derived from the patient's own tumor tissue. As Coffin points out, the poster child for this approach is GVAX. (See Pardoll, Chapter 8.)

"So, what we thought was to combine the two. Combine an oncolytic, local killing approach with a vaccination approach that has systemic efficacy by generating an autologous vaccine directly in the patient."

To accomplish this, you engineer your herpes virus to express a drug that invigorates the immune system (in this case, the same granulocyte-macrophage colony-stimulating factor [GM-CSF] used in GVAX), and then inject the virus directly into the tumor. Infected tumor cells then rupture, releasing their mutated innards (including tumor antigens) into the surrounding tissue where the immune system's dendritic cells—recruited by the GM-CSF—clean up the mess and then give orders to the rest of immune system to scour the entire body for any cells that contain any of those antigens and if found, kill them.

That's T-Vec: An oncolytic, personalized vaccine.

"You don't have any external processing like other autologous vaccines," explains Coffin, "You just inject the virus into someone's tumor, it melts the tumor, and [the antigens and dendritic cells] make it into a vaccine in situ." In this way, and unlike standard vaccines, the components of the vaccine are determined by the patient's immune system. "You don't need to know anything about the patient's tumor," says Coffin, "T-Vec essentially provides a sort of universal tumor vaccination." This capacity has been clearly demonstrated in clinical trials where metastatic tumors—tumors that were not injected directly with T-Vec, that were not infected with HSV—went away. Only a vaccine-like effect would do that.

> *"You don't need to know anything about the patient's tumor."*

On the basis of these robust clinical trial outcomes, T-Vec was approved in 2015 for the treatment of melanoma.

Coffin's new company, Replimune, is at work optimizing the HSV platform for use in other cancer settings.

✤ ✤ ✤

Favorite science book?

"I would point people in the direction of *The Selfish Gene* and *The God Delusion*, both by Richard Dawkins." The latter of the two choices is obviously philosophical. "I don't think there's a place for religiosity in a scientist. Science is supposed to be rational."

And yet, we have Francis Collins, famous scientist and evangelical Christian who headed up the Human Genome Project and is currently the head of the NIH. "We do indeed. I don't know quite how that works, but I would suggest to everybody they should read *The God Delusion*."

On the strictly science front, there is *The Selfish Gene*, a treatise on evolution. "It's the key concept in biology," says Coffin. "It underlies everything, and it's a very simple concept to get your head around. If you think about everything in evolutionary terms you get yourself a very long way in biology and in cancer research as well."

Endnote: In response to Dr. Coffin's pondering on how science and religion could possible mix, there is a book that enlarges the conversation: The Language of God: A Scientist Presents Evidence for Belief, *by Dr. Francis Collins.*

T REGULATORY CELLS (Tregs)

Treg Cells Control a Variety of Immune Responses

Bone Marrow

Thymus

Foxp3⁺

Treg — CD25⁺ Foxp3⁺ CTLA4⁺

Naive T cell

Foxp3⁺

Effector T Cell

Autoimmune Disease
Allergy
Inflammatory Bowel Disease
Tumor Immunity
Organ Transplantation
Feto-Maternal Tolerance
Immuno-Metabolic Disease, etc.

Dec. 6, 2016

Shimon Sakaguchi

SHIMON SAKAGUCHI
"What Tregs Do"

CHAPTER TWENTY

Shimon Sakaguchi, M.D., Ph.D.

Distinguished Professor
Department of Experimental Immunology
Osaka University
Osaka, Japan

"T-cell biology all collapsed and everybody walked away from this kind of suppressor T-cell research. That was kind of sad, but it happened."

—S. SAKAGUCHI

Shimon Sakaguchi was born in 1951 in a rural area of the Shiga prefecture, an hour east by train from the city of Kyoto.

"My mother is still there," he says, "so for me, it is a special place." Sakaguchi's mother comes from a family of village doctors, and little Shimon was expected to walk a similar path. However, this legacy was at odds with Sakaguchi's interests, which more closely resembled those of his father, a high school teacher who majored in philosophy and spoke French. "My father was teaching humanities, so I preferred to study something related to humanities ... literature, art, something like that." In fact, as a child Sakaguchi enjoyed drawing and dreamed of one day becoming an artist. "Maybe a painter ... something romantic."

Were it not for global warfare that dream might have come true, but the only thing Sakaguchi's father's humanities cred got him—in particular, his knowledge of French—was a ticket to the front lines in French Indochina. Having survived the carnage of those battles, all the senior Sakaguchi could think of after the birth of his son was how to keep him out of the next war, concluding that a training in the natural sciences, if not medicine, would help his son to avoid a military draft or at least keep him out of combat. "So I chose a kind of compromise," says Sakaguchi. Yes, he would become a doctor, but maybe a more interesting kind of doctor, like a psychologist, "Or maybe psychopathology ... something like this. So then my father is happy and also maybe my mother is happy too."

Off to medical school he went. The interest in psychology eventually fell away, but rising up in its stead was the largely unexplored (at the time) puzzle box of immunology. This is not to say that Sakaguchi lost interest in the humanities. There was still art, and there was philosophy.

"Let me explain to you this way. So, my hobby is visiting museums," a habit Sakaguchi indulges lavishly as he travels the world giving lectures. He doesn't do this because he considers himself to be any kind of art maven, but rather because he has an acute understanding of what it's like to see something differently and to expand the world of knowledge through that unique perspective. "The artist is adding something original. Even if he is looking on the same landscape as someone else, he sees it differently," Sakaguchi says, and science is no different. "What we are looking at, it will be different depending on the scientist, and that is something very common between art and science, scientist and artist, [which] is something fascinating to me. I visit museums, and look at the paintings and then imagine how he or she painted this, how she or he looked at, say, a still life, or landscape, or another person." If it truly is art, if it is the fruition of a cultivated science, then right there in front of you is something new, something that no human being has ever seen.

As for philosophy, "Certainly I was influenced by my father's way of thinking. In immunology, there is something philosophical there," something quite metaphysical, and to Sakaguchi, something old and familiar: the balance of yin and yang or, in immunological terms, the phenomenon of self versus non-self discrimination. "As a medical student this was something fascinating to me," says Sakaguchi. "First we are studying how the body is protected from invading pathogenic microbes by the immune system, but then we look at diseases where the immune system attacks our own tissue."

At first glance, these opposing forces seemed perverse, but Sakaguchi soon learned of other examples. "If you have a wound, the blood should coagulate in that place and stop the bleeding. That's okay. But if the same thing happens inside the vessels—coagulation of blood, clotting—it's a big problem. It's a dichotomy: good and bad. But how it can be done, how can it be regulated, and what's the mechanism behind this?" In asking these questions, Sakaguchi smiles like an explorer with a brand new compass. "The mechanism should not be a simple type of thing. No, it must be something exquisite." It just has to be. He knew it. "And

> *The mechanism should not be a simple type of thing. No, it must be something exquisite.*

that was my start in immunology." Shimon Sakaguchi earned an M.D. in 1976, and a Ph.D. in immunology in 1982 at Kyoto University.

Ladies and Gentleman: The Tregs!

Tregs. It sounds like a good name for a band: The Tregs! But Tregs (pronounced tee-regs) stands for regulatory T cells. When Sakaguchi discovered them, he was not thinking even a little bit about cancer; his focus was on auto-immune disorders and whatever might keep a person from getting them. "I got interested in a very peculiar finding by a Japanese scientist," Sakaguchi recalls. It was the result of a very peculiar experiment. "What they did is removed the thymus from mice on day 3 after birth, and what they find is that the ovaries are destroyed." If they removed the mouse thymus at day 7, the ovaries developed normally.

The members of the group doing that work were endocrinologists by training, so they figured the effect resulted from the loss of some kind of hormone produced by the thymus. Sakaguchi, the immunologist, thought otherwise; he thought the loss of ovaries was an artifact of an induced auto-immune disease.

"It was the early 1970s, so from an immunological point of view the thymus still was a kind of fascinating organ," Sakaguchi explains: "It's producing lymphocytes, right? Removing the thymus means somehow the lymphocytes are affected and cause autoimmunity. To explain this, we came up with [the concept of] regulatory T cells" (i.e., a cell that resides in the whole population of lymphocytes that suppresses the immune response). Hypothesis in hand, Sakaguchi then had to provide evidence for, and defense of, what was later perceived as a flawed, if not embarrassing idea.

In Pursuit of the Exquisite

Exploring his hypothesis was a lot of work, with most of it done at the worst possible time for the suppressor T-cell field (see below). The logical starting point was to remove the thymus from the puzzle box; Sakaguchi was certain it did not belong in the big picture. It was a simple experiment: Remove the thymus from a newborn mouse as well as all the T cells, and then replace the removed T cells with a transplant of T cells from a genetically identical, intact mouse. The result? The no-thymus, no-T-cell mouse with the T-cell transplant from a regular mouse did not develop auto-immune disease, and it did not die.

The next order of science was to determine which subset of T cells was responsible for this mouse rescue. In general, T cells come in two types: CD4-positive (CD4$^+$) or CD8-positive (CD8$^+$). It was already known that CD8$^+$ cells are so-called "killer" T cells and have little to do with T-cell suppression. Therefore, Sakaguchi started looking at cells that were CD4$^+$. "We used various cell surface molecules to dissect this population," he explains.

Sakaguchi began by making a lymphocyte suspension from the spleen of a normal mouse (mammals keep a store of lymphocytes in the spleen). That suspension was then passed through a filtering process using an antibody to a T-cell surface molecule called CD5 (nobody's quite sure what CD5 actually does, but it is a useful tag nonetheless). It was hoped that this process would result in a test tube of T cells minus all the T-cell suppressors. This preparation was then transferred into a mouse that had no T cells of its own. The result: The mouse that had no cells that suppress the immune response developed multiple autoimmune diseases within days and died. "We predicted that these mice [would] spontaneously develop autoimmune disease, and indeed it happened," Sakaguchi notes, but he also knew that his approach probably removed more than just Tregs. "So the next question is, how we can narrow down to a specific population?"

Concurrent with this work, another group at Oxford University used an antibody to a molecule called CD45RC (a known type of tyrosine phosphatase enzyme) to perform a similar cell separation and achieved similar results: The transplant of suppressor-depleted T cells killed their mouse. "So I remove suppressor cells one way, and they remove suppressor cells another way. So this is what we are looking for: cells with CD45RC and CD5." However, using both markers as a method of cell separation was too cumbersome. Sakaguchi looked for a simpler way. That way was to focus on CD25, a marker of activated T cells. "CD25 is just perfect because it selects for cells that have CD45RC *and* CD5," says Sakaguchi. "Using this, we removed that population and the results were fantastic. The mice got spontaneously autoimmune." Suppressor T cells did exist, and Sakaguchi knew how to find them. "In 1995 we [showed] that CD25 is the best marker for this population. This is the most cited paper in *Journal of Immunology*, even now."

There is no little irony in this achievement, because in 1995 the editor of the *Journal of Immunology* was Ethan Shevach, who hated the idea of Tregs conceptually, and it was Shevach's laboratory that had produced the CD25 antibody (albeit for reasons having nothing to do with suppressor cells). "So when Ethan sees my work he tells his postdoc, 'You know, I don't think this work is any good. How about you repeat his experiment.' So she does the experiment, and she gets the same result." And just like that, Ethan Shevach not only published Sakaguchi's work, he became one of Tregs' most prominent advocates.

Deep Dark Night

The work that culminated with Sakaguchi's aha moment was performed over a roughly 15-year period. During the middle of those years, the suppressor T-cell field lost all credibility.

"I have a story," says Sakaguchi, with a quiet sigh. The story is highly technical and involves at least one very prominent scientist who, for reasons of professional courtesy and respect, shall not be named here. Briefly, a group of scientists injected T cells into mice and one of the results was the production of an antibody that was called anti-I-J determinant. The name was given because the antibody seemed to bind to things that were suspected to be part of the T-cell suppressor mechanism and, as such, was an extremely useful tool to study these cells. Yet, the technology of the time was comparatively crude, so much of the information regarding the location and function of I-J was based on indirect observations. Nevertheless, the authenticity of I-J was widely accepted, and was the foundational concept for many investigations regarding suppressor T cells.

So, all was well. Basic science was flowing, and there was a reliable stream of related funding for the field. Then, in the early 1980s, technical science caught up. Techniques to study cells at the molecular and genetic level came on line, and some of the discoveries resulting from these techniques were eye-opening. In 1983, a paper written by Leroy Hood, who revolutionized gene-sequencing technologies, stated that the proposed genetic location of the I-J determinant did not, in fact, actually exist.

It was an academic disaster. The support for research in suppressor T cells evaporated overnight. Grants were lost, labs were lost, and investigators in prominent academic positions were demoted. "T-cell biology all collapsed and everybody walked away from this kind of suppressor T-cell research," Sakaguchi says. "That was kind of sad, but it happened."

And that wasn't all; there was piling on. Sakaguchi's T-cell observations came into question. "People—even some Nobel laureates—said, 'Well, maybe when you removed the lymphocytes it caused an immune deficiency, and so the mice got infections and that is what caused the autoimmunity.' People thought this way," Sakaguchi recalls. "So, that was not so good for me."

Nevertheless, Sakaguchi was able to continue his work. "I had looked for scholarships and very fortunately was given the Lucille P. Markey award." The award was both prestigious—one has to be nominated by one's institution—and it was generous. "It supports you for eight years of postdoc, and then the transition to faculty, so I was very lucky." The award brought Sakaguchi to the United States and helped to support him through his scientific journey. Stops on the way included Johns Hopkins, Stanford University, and the Scripps Institute, with an eventual return to Japan and a faculty position at Kyoto University.

The Bomb

The Treg story, at least for our purposes here, is nearly finished, but first a few words about the atomic bomb. The bomb made two significant contributions to science with regard to Tregs. First, the bomb brought Sakaguchi's father back home, alive and well, where he was instrumental in advising his son, the future scientist. Second, the bomb produced the so-called "scurfy" mouse, which was created at the Oak Ridge National Laboratory in Tennessee.

"That's a story," says Sakaguchi. "In World War II the atomic bomb is dropped on Hiroshima and Nagasaki and then after that the Cold War happens." At the time, it was surmised that a nuclear exchange between superpowers was entirely possible, if not likely. "The United States government, they must study the effects of the radiation on mammals, so at the National Laboratory they are making mutant mice by irradiating them and picking out mutants. One of them is scurfy mice." Scurfy mice have symptoms of autoimmune disease.

Leveraging modern genetics technology, scientists were quickly able to track the gene responsible: FoxP3. "But nobody knew how this gene causes autoimmunity." They did, however, find that the genetic mutation was also seen in patients with a disease called IPEX syndrome, the symptoms of which reminded Sakaguchi of mice lacking Tregs. In short order, Sakaguchi's group demonstrated that FoxP3 is a transcription factor for Tregs and, moreover, that FoxP3 is the master regulator of Treg development. In fact, if you throw the FoxP3 switch in naïve T cells (i.e., those that have never been activated to perform a function), those T cells will become Tregs. The identification of FoxP3 is a fundamental discovery in T-cell biology, and Sakaguchi's paper on the subject was published in the premier journal, *Science*, in 2003 (Hori et al., *Science* 299: 1057 [2003]).

Cancer

But wait—this book is about cancer, not autoimmune disease—yes? "Well, to me it's the same problem," offers Sakaguchi. "The cancer antigen is a kind of self-antigen, or quasi-self-antigen." A tumor is just a mutated form of you, and T cells are greatly discouraged from going after what is you. If you get cancer and your T cells don't respond with the required intensity, Tregs often have everything to do with that. Theoretically, they are trying to protect you. "All the same principles apply." For the last six years, Sakaguchi has been applying those principles to cancer immunotherapy research in his lab at Osaka University.

The idea of cancer immunotherapy might well have rescued Sakaguchi's career. "I must tell you something," Sakaguchi says quickly, and then pauses.

"When I found the regulatory T cells and then showed it in the first paper using mice, and that maybe it can be used for tumor immunity, immediately, Lloyd Old from Memorial Sloan Kettering, he got interested in this work. Actually he called me to his meeting, this Cancer Research Institute meeting and I gave a talk." By that time, the existence of Tregs was more or less accepted, but not particularly celebrated, as the reputation of the field had been so badly damaged. Nevertheless, Lloyd Old, a man with an uncanny ability to spot the right idea, was fully on board, so much so that in 2004 he bestowed on Dr. Sakaguchi and fellow Treg proponent Ethan Shevach the coveted William B. Coley Award for Distinguished Research in Basic and Tumor Immunology.

The award went a long way toward healing some deeply felt hurts. "I really appreciate Lloyd Old," Sakaguchi says, struggling a bit to maintain his scientist's composure. "He was like a grandfather figure to me. He gave me this prestigious award, and then people started thinking that, well, what this guy is doing, these regulatory T cells, — this

> *"I really appreciate Lloyd Old. He was like a grandfather figure to me."*

is maybe something for cancer immunotherapy." It was all Dr. Old's doing. "Only three times I met him here in the United States, always at CRI meetings," Sakaguchi says with sustained wonder, "But he recommended me to the National Academy of Sciences in the United States. It is very difficult to become a member of the National Academy of Sciences here from [the] outside, five or 10 times more difficult than maybe for United States scientists, but he recommended me for this."

Dr. Shimon Sakaguchi was elected Foreign Associate of the National Academy of Sciences in May of 2012.

Sakaguchi's champion, Dr. Lloyd Old, died in November of the preceding year.

✤ ✤ ✤

Academic recognition was not the only reward for Sakaguchi's years of work. Drugs targeting Tregs are currently undergoing clinical testing in multiple tumor types including cancers of the pancreas, lung, skin, and ovaries, as well as leukemia.

Zen and the Art of Science

What's the worst thing about being a scientist?
"The worst thing? It is very addictive. You cannot stop."
What's the best thing about being a scientist?
"The best thing …" Sakaguchi considers for a moment before he pegs the answer: "The best thing is that I do not want to stop!"

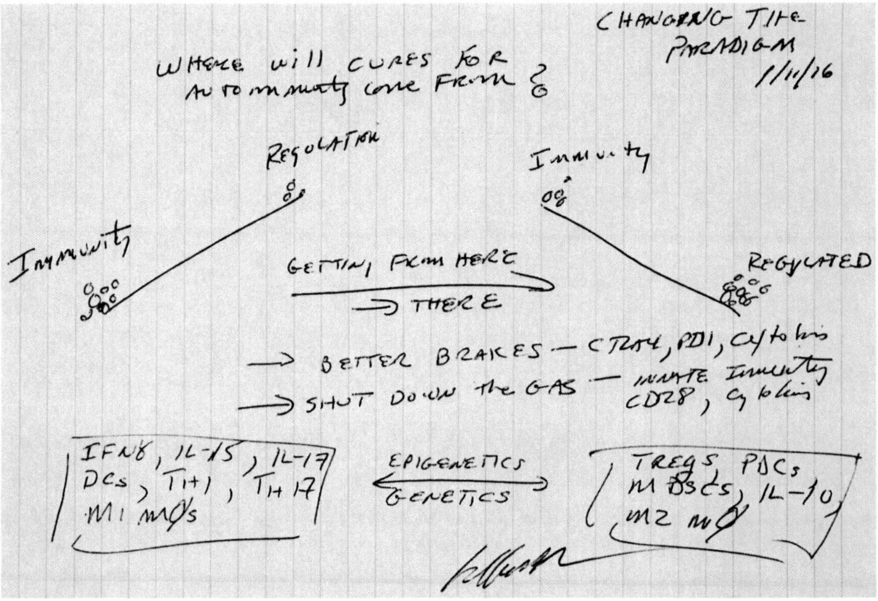

JEFF BLUESTONE

"Changing the Paradigm"

CHAPTER TWENTY-ONE

Jeff Bluestone, Ph.D.

Clausen Distinguished Professor of Metabolism
and Endocrinology
President and CEO of the Parker Institute
for Cancer Immunotherapy
University of California
San Francisco, California

"This is cool. I gotta work on this." —J. BLUESTONE

Jeff Bluestone is 64 and originally from New Jersey. "You know Exit 131 on the Parkway?" Everybody in New Jersey identifies themselves by their exit. "I spent my formative years in Colonia, then Edison," says Bluestone, at Exit 131: Metuchen. "Edison is known for Thomas Alva Edison in Metuchen which—it's funny—it's like a doughnut and the hole is Metuchen, the township, and Edison is around it." Edison's original lab is still there. "It's a famous building with a light bulb on it."

Despite the obvious symbolism, Bluestone did not have any particularly bright ideas early on. He was a typical kid. He had a rock collection: Rocks were cool. Science was cool in a general sense but it was not, as he puts it, "a transformational thing." There was a lack of focus, to be sure. "My parents took me to one of those counselors, you know … They told me I should be an orthodontist or something like that. And then I was going to be a veterinarian, because I thought 'Veterinarians … that's kind of cool. I can do science on animals and that sort of thing.'" Bluestone is pretty laid back, giving the impression of a man just a bit spent, but still resonating from serious fun, like a retired surfer. One can easily imagine him padding about the lab in Crocs and a Hawaiian shirt.

Not too far from Exit 131 is Bluestone's first alma mater, the New Brunswick campus of Rutgers University where, in his fourth year, he finally heard the call. "Yeah, senior year," remembers Bluestone, "That was the epiphany." That was the year Bluestone signed on to do a research project with a man named Bob Cousins, who was then a junior faculty member.

Sometimes mentors find you. Sometimes you find them. Bluestone sought Cousins out. "Yeah, he gave a lecture in one of my classes and I said to myself, this guy is a great guy, so I went to him and I said, 'Can I just volunteer in your lab?'"

The attraction for Bluestone was more personal than scientific: Cousins had a passion. "My whole family is insurance salesmen," says Bluestone. "Nobody in my family ever did science, so this was the perfect fit for me to have somebody like that ... somebody who saw the opportunity and had the passion. Especially when you're a little older like I was and you don't know what you're doing, you need somebody who's going to give you that spark, to light that fire underneath you. A mentor is key to that."

Up to that point, most of Bluestone's academic experience had been a repetitive pain. "It was like, I have to take organic chemistry, I have to memorize all these molecules ... It just didn't have that creativity. Classes are not creative. People are creative."

And sometimes people are really lucky.

"Bob was working on a calcium-binding protein, which seemed kind of ridiculous to me," says Bluestone. "I was in the Ag school ... he was in the Ag school as a nutrition guy. The bottom line was that that little calcium-binding protein he was working on turned out to also be a *calmodulin*-binding protein, which turned out to be calcineurin, an incredibly important molecule."

Calcineurin, it turns out, is an activator of T cells. Calcineurin inhibitors, therefore, block T-cell activation, which is critical, if not lifesaving, in the setting of allogeneic (i.e., foreign tissue) transplantation (see GvHD, Chapter 8). By facilitating organ and bone marrow transplants, calcineurin inhibitors have saved tens of thousands of lives.

"So I got very excited about science then, not as a hobby but as something I wanted to do with my life."

The Primacy of Pondering

Any scientist who speaks of absolutes is a bad scientist. Certainty is the signature of a fool.

"I heard a lecture a while ago by a guy named Paul Nurse, a Nobel Prize winner [2001] and the head of the Rockefeller University," says Bluestone, "And he talked about doubt." What actually makes scientists unique is that they don't accept anything at first glance, or second glance, or even third. "There's always this doubt ... do we really know everything about it? Do we really fully understand?"

A good scientist does not say, "Based on my work, so and so is true." Instead, they say, "Based on my work the data *suggest*." That's it. It's a suggestion. It's hypothesis generating. The data never flat-out prove or say anything. The data suggest that perhaps, taking all things into consideration, one possible interpretation could be ... It might read like waffling, but scientific truth is rarely, if ever, a sound bite.

At the time Nurse gave his lecture, the issue of certainty was the stuff of international debate. It was the early 2000s: 9/11 was still fresh, and bin Laden was Public Enemy No. 1, and ... *Saddam Hussein had weapons of mass destruction! Unquestionably!* "[Nurse] was a quite liberal guy," says Bluestone, "And he was talking about how the government in the United States [Nurse is English], about how there was no doubt ... whatsoever that Saddam Hussein had weapons of mass destruction, and he was saying, that's the difference between politicians and scientists: scientists question everything." *Note: No weapons of mass destruction were ever found.*

Merely being a skeptic is not enough, of course. "Curiosity is important. Imagination probably is important, because you have to imagine what it could be if it's not X. But to me it's probably more about just saying, 'Do we really think we know everything?' The answer is always, 'No.'"

"There's a book I read a few years ago called *The Beginning of Infinity* [by David Deutsch], and this idea that every time you think you know everything you realize how little you know. To me that questioning of things and wondering what is the reality ... It's that doubt that's been a driver for me."

IO and CTLA-4

There's doubt, and then there's just being at a loss. When Bluestone began his Ph.D. at Cornell in the late 1970s, the field of immunology was just that: a free-range space full of unnamed things. "Yeah, immunology was phenomenology and it was crazy and it was exactly what I wanted to do," says Bluestone. He'd already had inklings about the immune system while working on his Master's degree in virology, but it was that very work with viruses that spurred him to aim higher.

"What I did for my project was ridiculously small at the time. We had a virus—mengovirus—and my job was to figure out how many adenosines were hooked onto the end of the mengovirus," which would affect its translation into protein. It was a small question from Bluestone's perspective, and the clinical utility, if any, was not obvious. "I know it probably mattered to somebody, but I wanted something that was going to be ..." Bluestone wanted to make a difference, "And to me, working on cancer would do that."

Contributing to cancer required a change of venue. "I spent a summer at Sloan Kettering working with Bob Good [one of the founders of modern immunology]," says Bluestone. "It was so complicated, and so challenging to figure out how the immune system was going to impact on cancer that I just knew, I said to myself, 'This is cool. I gotta work on this.'" That desire led Bluestone to a benchtop at the National Institutes of Health (NIH). "In the 1980s I was a T-cell guy working down the hall from Steve Rosenberg (Chapter 13). We were cloning T cells. He was doing it in humans; I was doing it in mice."

One of the more important aspects of the work was to isolate and clone (i.e., reliably reproduce) the T-cell receptor: the business end of the T cell that recognizes and binds to the enemy antigen. Once cloned, scientists would be able to unravel the intricacies of its function. "Mark Davis [of Stanford] eventually cloned the mouse version and it was great," says Bluestone. Having such a resource in hand freed him up to ask a multitude of questions and, in the asking, reveal many more.

"We came across an observation which was exploited by Drew [Pardoll, Chapter 8]. He and I published a *Nature* paper on it in 1987 [Bluestone et al., *Nature* 326: 82 [1987]]." They had identified an antibody that reacted with the cloned T-cell receptor complex, but the antibody interaction, which should have behaved like a foreign antigen and activated the T cells, did not do so. "It was not sufficient," Bluestone explained, "We couldn't get the T cells activated with the antibody alone. You needed something else: a second signal."

The second signal was a molecule called CD28, which had already been isolated in other labs but without its receptor-binding partner (signals always bind to receptors). "In the late '80s, Ron Schwartz [at the NIH] and Marc Jenkins [University of Minnesota] isolated the human version, and Carl June [Chapter 16] and Craig Thompson did it in mice," recalls Bluestone, yet no one could figure out what the CD28 molecule bound to.

There was this, however, a researcher named Peter Linsley (of the Benaroya Research Institute), who was working with a gene called CTLA-4, which was initially derived from a cytolytic T-cell (CTL) that had been grown in a culture plate with numerous tiny wells laid out in a grid, allowing for multiple experiments to be run simultaneously. The molecule of interest resided in that plate's well at the intersection of row A and column 4: well A4. Thus, CTLA-4.

The figures were all lined up just waiting for someone to do the math. "It was Craig Thompson … no, it was Craig Thompson's wife, Tullia [Tullia Lindsten, M.D., Ph.D., at MSKCC]. She was at a meeting and looked at this thing and noticed that it looked a lot like CD28." This is not surprising, in retrospect, as both molecules are receptors and both receptors bind the same signal. Tullia relayed this observed similarity to Peter Linsley. "So Peter took the *CTLA-4* [gene] and made something called CTLA-4-Ig, which was a soluble form of CTLA-4," says Bluestone, and this fusion protein made for a good probe. "And I was at a conference struggling with this whole thing and Peter gives a talk and says, 'I've made this molecule CTLA-4-Ig and it binds to cells, antigen-presenting cells, and it looks like CD28.' And I'm like, 'Aha, I'll bet you *that's* the second signal that I can't understand!'"

CTLA-4

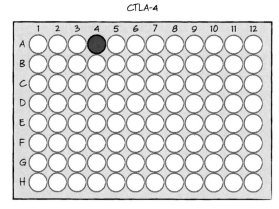

After the talk, Bluestone approached Linsley. "I say to him, 'Peter, I've got this great system. We're doing these transplants of …. these human islets into mice and I can study the second signal. Can I get some of that stuff?'" A week later, the material arrived at the Bluestone lab. "The world was different in those days," notes Bluestone of the exchange. Back then, the sharing of materials between an academic and a company man was not subject to so many permissions. "But there was no MTA, no nothing."

Note: MTAs (Material Transfer Agreements) are legal documents regarding the rights and responsibilities of parties that are exchanging research materials. The purpose of an MTA is to make the initiation of scientific collaborations both time-consuming and unpleasant.

Islet cells are the cells in your pancreas that produce insulin. Islet cell transplantation is proposed as a way to treat type 1 diabetes, an autoimmune disorder in which your immune system mistakenly wipes out all your islet cells, thereby making you insulin-dependent for life.

Bluestone put Linsley's construct to immediate good use. "I walked into the lab and I said to Debbie Lenschow, who was my M.D., Ph.D. student at the time, I said, 'Debbie, do me a favor. Just take this stuff and give 50 mg every other day to these mice and see what happens.'" Lenschow did, and the islet cell–transplanted mice were no longer at risk from rejecting those transplants. The immune system was somehow turned off by Linsley's stuff.

Note: An update on Peter Linsley's stuff: Turning off the immune system with CTLA-4-Ig was not merely a nifty scientific trick. After further consideration and testing, "Linsley's stuff," now known by the brand name Orencia, was approved in 2011 to treat the autoimmune disease rheumatoid arthritis. Sales for Orencia in 2015 topped $1.8 billion.

"So I said, 'Ah! We've found how CD28 works.'" The thought process went like this: T cells require two signals to activate. Signal one is delivered when the T cell encounters its antigen. Because CTLA-4-Ig (i.e., the soluble CD28 look-alike) stuck to dendritic cells (DCs; see Steinman, Chapter 10) in solution, it was determined that the second signal most likely came from the DCs. This suggested that there was a signal coming from the DCs that bound to the CTLA-4 receptor, and that flooding the system with CTLA-4-Ig somehow blocked that second (DC-bound) signal, thereby preventing it from completing the activation of the T cell.

Ultimately, the model Bluestone posited was this: T cells get their first signal when they encounter an antigen, but don't fully activate until molecules on the T-cell surface (the B7 molecules) attach to a receptor molecule (CD28) on the DC. Here's the tricky bit, though: CD28 and CTLA-4 look very similar to those B7 molecules, and the B7s can also bind to CTLA-4. Bluestone's experiments suggested that the T cells could be fooled by "fake" CD28 (i.e., CTLA-4-Ig). By flooding the system with soluble CTLA-4-Ig, the B7 molecules on the T cells preferentially bound to that, instead of to the CD28 on the DCs. Without the stimulatory effects of CD28, the T cells failed to activate fully, and there was no transplant rejection. The conclusions from these experiments were published in the journal, *Science*, in 1992 (Lenschow et al., *Science* 257: 789 [1992]).

At this point, the entire Bluestone team started working on the CTLA-4 pathway in an attempt to nail down its function. Graduate student Theresa Walunas (now an Assistant Professor at Northwestern University) was charged with immunizing hamsters with CTLA-4-Ig to make **antibodies** against regular CTLA-4 that could be used to block *its* activity. Hopefully, that loss of activity could be readily observed.

*To make an **antibody**, first purify the protein of interest. Inject that protein into an animal (e.g., hamsters, rabbits) to which that protein is foreign, and wait for the animal to mount an immune response by generating its own antibodies to attack it. Those antibodies can be purified from the animal's blood, are exquisitely tailored to bind strongly to the protein injected initially, and can be used in any number of ways as a tool to explore the protein's function.*

When Walunas performed the experiment with her new CTLA-4 antibodies, the results were indeed obvious: T cells were activated. Concurrent with this work, other labs chasing the same prize also suggested that CTLA-4's effect on T cells was stimulatory. This was interesting, but did not serve the overarching needs of Bluestone's research. "We wanted inhibitors, not activators." He was doing transplants and wanted a way to shut down

T-cell-mediated rejection. "So I said, 'Theresa, do me favor, take the antibody and make Fab fragments of it.'"

In general, if an intact antibody is pared down to only the *fragment* that recognizes and that binds to its target (hence, Fragment, antigen-binding [Fab]), then all that should happen is that the binding pocket is blocked and the function associated with that molecule/receptor interaction does not occur. In this case, CTLA-4 with Fab bound to it should not activate.

Walunas made the Fab and applied it. T cells were activated. "And I said, 'No, no, no. Fabs don't activate. Fabs inhibit.' We were sitting there at the blackboard, and I said, you know, maybe the molecule doesn't do what we think it does. Maybe it's not an activator. Maybe it's an inhibitor. And maybe by using the Fab we were inhibiting the inhibitor and that's why we got a boosted response."

It was a lot of maybes. "But that was the day—Theresa and I talked about this many years later—that was the day we came to that conclusion," says Bluestone: CTLA-4's function was to inhibit the activity of T cells.

The challenge after that was to convince someone—anyone—that their conclusion was valid; remember, other respected labs thought CTLA-4 was *stimulatory*. "We submitted the paper to *Science* and they turned it down, then *Nature*, and they turned it down. Nobody believed that there was such a thing as a negative regulator on a T cell."

With the help of a personal friend, Laurie Glimcher (at Dana-Farber) who had just started a new journal, Bluestone's work was published in 1994 in the journal *Immunity* (Walunas et al., *Immunity* 1: 405 [1994]).

"Jim [Allison] published his paper about a year later with Max Krummel [at UCSF] replicating what we had done," says Bluestone, and then roughly a year later, working with Arlene Sharpe (at Harvard) a CTLA-4 knockout mouse died within two weeks of birth from massive autoimmune disease.

Turn it On

If you turn CTLA-4 off, you can treat cancer. If you turn CTLA-4 on, you can treat type 1 diabetes. "Everything sort of builds on something else, right?" That's science for you, says Bluestone. From the same starting point, one can venture forth on very different roads (with very different difficulties). Using an antibody (ipi) to attenuate CTLA-4 activity? Simple enough. Trying to selectively enhance that activity to deactivate T cells that are attacking the cells of the pancreas? *Eeesh.*

"It's backwards, right? But [activating] is much harder to do than to block the negative regulator," says Bluestone. To succeed, first he needed more information; he needed to drill down on how CTLA-4 exerts its effect. "So we started looking where the molecule was expressed and we found this very small population of cells that constitutively expressed the molecule." These cells are T-regulatory cells, or Tregs. "I thought, well since CTLA-4 is essential—you die in two weeks if you get rid of the molecule—then Tregs must be essential."

Tregs had been identified previously (see Sakaguchi, Chapter 20), although their mechanism of action was unclear. What was known was that they expressed both CTLA-4 and CD28. To dig deeper, Bluestone instructed one of his postdocs to look at the fate of Tregs in mice that had both CTLA-4 and CD28 knocked out. Further, Bluestone required that the mice used be genetically susceptible to autoimmune disease.

"We chose this mouse that gets diabetes," explains Bluestone, "And it turned out that the animal got diabetes within days after you eliminated these molecules, and the only thing that was missing in these mice was this small population of cells." The conclusion was that Tregs somehow prevent type 1 diabetes.

"So I switched 80% of my lab over to studying Tregs," the purpose being to work on clinical translation. "I decided that we're going to do this differently," says Bluestone. "Instead of making a drug, we're going to make cells into a drug." Bluestone's idea was to take Tregs from organ-transplant patients and from patients with autoimmune diseases, expand the population of those cells in a dish, and then inject them back into the patient with the hope of overwhelming the errant immune response. At the time of this writing, there are at least six such clinical trials underway, as well as several others specifically addressing GvHD in the setting of allogeneic stem-cell transplant.

Another Treg immunotherapy approach under investigation is specific to type 1 diabetes. "The way to do that—and I've worked with a couple of companies on this—is to use embryonic stem cells, or iPS cells [see Sadelain, Chapter 17] to make insulin-producing islets." This method requires three things to be successful: (1) Get rid of the offending T cells that are destroying islets (there are drugs now that can do that); (2) inject new Tregs; and (3) replace the missing islets. "We've done all this in mouse models," says Bluestone. "We had a paper last year in *Cell Stem Cell* [Szot et al., *Cell Stem Cell* 16: 148 [2015]], in which we showed that you could actually induce tolerance in mice where you simultaneously give an islet transplant made from human embryonic stem cells and get [a] permanent cure from the diabetes that way."

The Dark Night

"Fear of failure is really a challenge, right?" Yet, no scientist in his or her right mind would expect to have every experiment, or even every other experiment, turn out the way you had in mind. The danger is not letting go of the loss. "I always tell students that in baseball if you bat .300 you can make it into the Hall of Fame. That means you failed 7 out of 10 times, right?" says Bluestone. "So that's why I never got too distressed over failure, whether it was not getting into a journal that I wanted, or whatever, I always knew that if it wasn't hard, it wasn't worth doing."

And if things do get heavy, he can always head into the light.

"Yeah, I got this neon sign that sits in the lab that my father-in-law made for me. It says: *Club Bluestone*. We turn down the lights, turn on the neon sign, and play Bruce Springsteen as loud as we can. And then we just keep going."

Another Jersey boy to the rescue.

Class I Shoulda Took in School

"Art history," says Bluestone, without hesitation.

Really?

"Look, science is incredible, but when you make the iPhone, what it looks like is as important as what it does. That's why Jobs worried about color, and fonts, and things like that. And me as a scientist, I don't often think about the aesthetics of what we're doing." Instead, science is an activity of reduction, a paring down of systems to their foundational core, thereby exposing the kernel of the phenomenon observed. Put simply, science is getting to know a whole hell of a lot about precious little. It's rarely a big picture pursuit unlike, say … making big pictures.

"Art history teaches you about aesthetics and about what matters," says Bluestone, "And I can tell you that you become much more astute, much more thoughtful about your science when you think of it in the context of the beauty. I

> *"You become much more astute, much more thoughtful about your science when you think of it in the context of the beauty."*

know this sounds crazy, but I think it's true. I would be so much better at my science if I could take it out of just being a bunch of numbers and a bunch of graphs and take it into a larger picture about the people that have been involved in it, about the patients, about the aesthetics of the experiment."

The minutiae, the small questions scientists ask can be so limiting, says Bluestone, but if he had taken art history, "I think I would be much more open to possibilities when I realize how these people, these artists, from the Man Rays to Picasso, how they could take a problem—the problem of how to illustrate what's going on in your mind—and come at it from so many different angles to get where they wanted to go."

Beauty is truth, truth beauty—that is all ye know on earth, and all ye need to know. —JOHN KEATS

CELLS AND SIGNALS: GOOD AND BAD

IDO/Treg loop

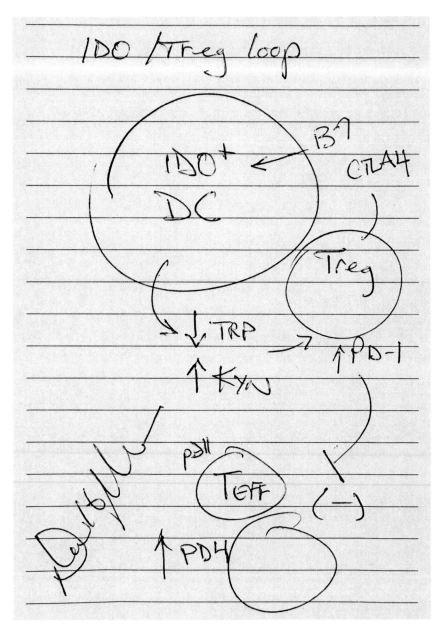

DAVID MUNN
"IDO/Treg Loop"

CHAPTER TWENTY-TWO

David Munn, M.D.

Cancer Immunology, Inflammation and Tolerance Program
Professor of Pediatrics
Augusta University
Augusta, Georgia

INDOLEAMINE 2,3-DIOXYGENASE (IDO)

"I think it was easier to assume that we were just crazy than to rewrite the textbooks." —D. MUNN

Note: Dr. Munn wishes the details of his private life to remain private. That wish is honored herein.

David Munn was born in Atlanta, Georgia, in 1957.

After undergraduate training in philosophy, Munn attained his M.D. at the Medical College of Georgia School of Medicine in 1984. Shortly thereafter, he discovered the focus of his life's work: childhood cancer. By the time Munn came to the pediatric game in the mid-1980s, medical science was finally putting therapeutic points on the board, and Munn wanted to run up the score. "I was interested in pediatric oncology mostly from a clinical standpoint," says Munn. "We'd gone from the 1960s, when virtually no pediatric malignancies were curable—it was a death sentence if you had cancer as a child—to the current situation where probably 80% of kids can be cured with chemotherapy." (He is speaking in particular here of pediatric hematologic cancers, not childhood cancers of the brain, where significant progress still awaits.)

Although impressive, those high success rates for treating pediatric leukemia/lymphoma came at a heavy price to the patients. "Those gains were purchased at a huge cost in toxicity," says Munn. Some of those toxicities emerge long after treatment has ended, often in heartbreaking ways. "These children have an increased risk of second malignancies in the future, they [have] school failure, [and] then there are all kinds of other subtle things." Regarding the vital parameter of cognition, one recent study found that IQ scores for children treated for acute lymphoblastic leukemia are lower by as much as eight points as compared to children that have never had cancer treatment. "It's just not a good thing to give high-dose chemotherapy to small children if you can help it," says Munn, "So to me this seemed like an area in which new research would be potentially high-impact."

> "It's just not a good thing to give high-dose chemotherapy to small children if you can help it ... so to me this seemed like an area in which new research would be potentially high-impact."

Once his post–medical school direction was determined—a fellowship at New York's Memorial Sloan Kettering under the guidance of Dr. Nai-Kong Cheung—cancer immunotherapy came quickly into view for Munn, though not precisely into focus. The reason for the fuzzy picture was conceptual. Bone marrow transplants for children had come along by that time, but it wasn't seen as a treatment for cancer, not directly anyway. "The idea that bone marrow stem-cell transplants were immunologic therapy hadn't really become clear in the early '80s. The goal of using transplants in pediatric treatment was primarily to rescue high-dose chemotherapy," because the chemo was wiping out not only the tumor, but the kids' immune systems, thereby leaving them highly vulnerable to life-threatening infections. The transplant was intended to reconstitute the patient's immune system. That's it. Transplants were *not* immunotherapy.

"I know that we were not thinking it was immunotherapy because at the time that I was doing my training the big thing was to deplete all of the T cells from the bone marrow graft so we would get totally pure stem cells," explains Munn. This protocol, designed to avoid the GvHD [graft-versus-host disease] side effect caused by T cells was developed at Sloan Kettering and at the Fred Hutchinson Cancer Research Center in Seattle, and given that young Dr. Munn saw these preeminent transplant centers doing their transplants that way, he figured it must be the right thing to do. "What they found out is that if you get rigorous 3-log depletion of all the donor T cells, you absolutely do get rid of graft-versus-host disease," says Munn. "The problem is that more kids die because they relapse."

When enough data from transplants were accrued, transplant investigators sifted through the numbers for an explanation. What the data told them was that for transplant patients overall, a little GvHD (it ranges in severity) was a good thing, because the graft itself was not just grappling with the host, GvHD was a sign that the graft (i.e., those donated T cells) was fighting the host's cancer. Clearly, patients with mild GvHD did better. "At that point, we realized that a lot of the success of bone marrow transplant was actually the immunotherapy effect," says Munn.

The observation stayed in the back of his mind. It was important, but there were too many aspects of basic science knowledge missing to explain the big picture phenomenon of this graft-versus-tumor effect. There were, however, some things known about the presumed causative agent, the T cell. "What really got me interested in immunotherapy was the fact that T cells were able to control viral infections, which is just a limited number of genes," Munn explains. "Some small viruses may only have 10 or 20 genes in them [people have around 20,000]. Yet, that is plenty for the immune system to recognize an infected cell and attack and clear it. And so, with all the hundreds and thousands of mutations that presumably are there in a cancer, why isn't the immune system attacking them? That was an interesting question."

Gung Ho, IO?

It *was* an interesting question, but at the time it didn't justify pursuing the topic as a profession. "When I went into immunotherapy the people who advise you about your career development said to stay away from immunotherapy." Certainly, work was being done at the National Institutes of Health (NIH) and elsewhere, but no one was even attempting to explain in any detail the specific effects on the physical systems that were impacted. The mechanisms, the details, the nitty-gritty of how it might actually work on a molecular level were unknown. If an agent was discovered to cause an observable immune response such as inflammation, it was given to patients.

> *"When I went into immunotherapy the people who advise you about your career development said stay away from immunotherapy."*

Results from these human experiments produced one measure of miracle for every nine shots of disaster. "This was the era of 'Let's try recombinant TNF [an immune effector] . . . Let's try recombinant interferon gamma . . . Let's push them to toxicity and hope that we can generate some inflammation before toxicity becomes too bad,'" recalls Munn. "And the answer was always that you'd just make people toxic with these things, and you don't get any added tumor effect."

The undisputed leader of these expeditions was Steve Rosenberg at the NIH (Chapter 13), a man with enormous investigative talents and clinical insights, as well as a gift for the art of soft-sell self-promotion. His work with the inflammation booster IL-2 produced both miracles and headlines. "Steve Rosenberg deserves a great deal of credit for that work," says Munn. He drove the field through a nasty conceptual roadblock: The widely held belief that that immunotherapy would not work because you can't make the immune system recognize a tumor. "He showed that if you were willing to put your patient in the ICU on a pressor drip, and third spacing, and losing whatever percentage of them he lost in the ICU on that [regimen], you could—in the ones that survived that—get regressions of immunogenic tumors." You could get cures.

The treatment was profoundly harsh, but there were a handful of home runs. The practical aspects were daunting, but Rosenberg proved it could be done. "Steve convincingly showed that the immune system was able to deal with even extensive disease if you pushed it hard enough." Unfortunately, the reason Steve had to push so hard was because the phenomenon of immune system suppression was entirely unknown, and it was that field-wide ignorance that kept the now roaring success of cancer immunotherapy at bay for many years.

It's not that inklings of the issue weren't out there. A number of investigators suggested that immune suppression mechanisms were an actual thing, but there was no proof. In fact, when Munn submitted his first papers about tumor immunology suggesting the existence of tolerogenic pathways in the immune system, he was roundly rebuffed. "The reviewers would say, 'you have to take that out; no one has ever shown that there is such a thing as an immunosuppressive pathway in the immune system.'"

Hey Baby

Munn's rebuttal to this learned denial was to simply ask: How do babies get born?

"When we wrote our first *Science* paper on pregnancy," says Munn, "The received opinion at the time was that the maternal immune system somehow didn't notice the eight pounds of baby that had been implanted into this woman's uterus, despite this implant having a complete hookup to the maternal blood system so that every T cell in her body flows through that placenta to the baby, a baby that has half of its genes from dad." And dad, in this context, is a very foreign genetic donor indeed. Yet, the understanding at the time was that the immune system remained ignorant of this interloper because there were no antigens presented.

It made absolutely no sense. "I was a pediatrician, so I knew that you had mixing of blood; you find fetal red blood cells in maternal circulation," Munn explains. "We knew that the mom was aware immunologically of this baby being there, but because there was no overt rejection response, the only hypothesis that could be formulated—given the scientific tools of that time —was that she must just not have noticed."

Yes, that must be it. For nine months, mom's immune system looks the other way while she gets as big as a house. It couldn't possibly be that mom's immune system can actually see the baby just fine, but that to preserve the human species, mom and spawn have somehow cut a deal so that mom's T cells will "tolerate" the baby. "But words like acquired peripheral tolerance—those three words—you weren't allowed to put together into a sentence when I wrote papers in the *Journal of Experimental Medicine*."

The Birth of
Indoleamine 2,3-Dioxygenase (IDO)

The IDO story begins with a phenomenological observation made during Munn's fellowship at Sloan, where he and his mentor, Nai-Kong Cheung, were co-culturing two types of cells from the immune system: T cells and macrophages. The goal of this work was to see if the macrophage, which is capable of engulfing and killing cancer cells, was also capable of activating T cells for the same purpose. The process is called "cross presentation," whereby an antigen-presenting cell (e.g., a macrophage) eats a tumor cell and "burps up" (i.e., presents) a molecular description of the tumor cell for the killer T cell to then go after. "But what we noticed when we put the macrophages and the T cells together was that the T cells really didn't activate; they were intensely suppressed." Mind you, not a lot was known about macrophages at this point, and dendritic cells, the cell type that actually should have been used in the experiment, had not yet been characterized by Ralph Steinman (Chapter 10).

"We didn't really realize that macrophages were not immunogenic antigen-presenting cells in the normal course of events for naïve T cells," admits Munn. That's what dendritic cells do. So, a bit of *"Oops"* there, but that little misunderstanding aside, the observation that macrophages were suppressing the activity of T cells was sound. "So, we were looking for what the mechanism of that might be, and one of the postdocs in the lab noted that the macrophages were metabolically active cells—that [was] certainly true and known—and he was concerned that there might just be a nutrient depletion problem." It was a co-culture system after all; maybe the macrophages were just hogging all the food.

To test this theory, the tireless postdoc (all postdocs are, by definition, tireless) stayed up all night feeding the cultures every hour with liquid nutrients to try and outstrip rampant macrophage consumption. Come the dawn, what the postdoc found was that under those overfed conditions the T cells activated just fine. The mystery was solved: The media was just spent. The lesson learned? Macrophages are pigs.

"That was a surprise to me because I thought that I had tested for a spent medium consumption problem," says Munn. He'd taken spent medium and doused a batch of T cells to see if they activated, and they did. "But, here's where my experimental design had been poor. I used 90% spent medium, and I didn't remix all my reagents in spent medium, and I didn't do my final wash of the T cells in spent medium." There was enough of *something* left in the remaining 10% medium that sent T cells in motion. It didn't seem reasonable, but it was so.

To find out what it was that the T cells liked so much, Munn's team used 100% spent medium and then added back each component one at a time. "It would be embarrassing to list the number of hypotheses that we confidently generated as to what the missing thing would be," says Munn. The first guess: iron. Iron is a key micronutrient, right? Well, it wasn't iron. Oh—well then—it's folate, right? Hm. Okay, not folate. Well then, it's glutamate, because glutamine (a metabolite of glutamate) is really important. Or, maybe not. "We went through this whole list, and, I don't know, maybe T comes near the end of the alphabet or something, but it wasn't until we bought each individual amino acid and added back a drop of each different one, and we went through the whole series and none of them had any effect, except when we added one drop of tryptophan and then *boom*, it completely restored the function of the T cells."

It was a simple, solid lead and Munn followed it. "Al Gore had only recently invented the Internet at that point, so you couldn't Google things like you do now," Munn points out, "But there was such a thing as a search engine, and that was Medline." Munn searched Medline with the key words "macrophage" and "tryptophan" and the engine produced five hits. "Five rather obscure papers, but definitely known papers that said macrophages consumed tryptophan by this enzyme called IDO." Beyond that, little about the function of this enzyme was known. Munn found that no knockout mouse had ever been created, and no one had ever tried to inhibit IDO's activity to see what would happen if they did.

> *"Al Gore had only recently invented the Internet at that point, so you couldn't Google things like you do now."*

Fortunately, Munn found that a good bit more was known about the function of tryptophan. "We were lucky. The tryptophan pathway had become of interest to the drug development world because it feeds into serotonin," and serotonergic agents, like Prozac, were making drug companies vast fortunes. Because of this (and with an eye on future fortunes to be made), a host of tryptophan derivatives had been created. Munn tested some of them in vitro. One of them blocked the activity of IDO.

The next logical step was to test the inhibitor in vivo, in a mouse. But what kind of mouse model would clearly demonstrate a response? "As we were tossing ideas around in the lab one of the postdocs said that this thing [IDO] was cloned from placenta, and placenta is certainly a good example of acquired peripheral tolerance."

This suggested a model. Based on this bit of information, Munn decided to create an antibody to the IDO enzyme, an antibody that could be used as an experimental searchlight that would show him exactly where in the placenta the IDO was residing, which would tell him more about its possible function. "It turned out that IDO was expressed specifically in the syncytiotrophoblasts, which are a very interesting cell type because they're the end of the fetus," says Munn. "They are the last fetally derived cell that comes in contact with the mother, and they are, in fact, the *only* cell type from the fetus that is in contact with the mother because they line the maternal blood spaces in the placenta."

It was all starting to make sense. For further testing, Munn chose a transplant mouse model, except in this experiment the transplant would be a mouse pup. It was an elegant choice. "It was great because nothing could be more dramatic than a sudden appearance of a fetus that's tolerated despite all of these clearly disparate transplantation antigens." If IDO, via tryptophan, was suppressing the mother's T cells so the baby could stay, then blocking IDO would cause the loss of the fetus, which would be a very obvious experimental result.

And so it was. With IDO blocked, the fetus was lost.

Munn worked to nail down the results: When the pup was the result of a mating where the mouse parents were genetically identical, blocking IDO did nothing. If the mating was between mouse parents where the dad differed by a single genetic change, blocking IDO led to miscarriage. The results were dramatic and the discovery profound, leading to a 1998 paper in the premier journal, *Science* (Munn et al., *Science* 281: 1191 [1998]).

And Cancer Is Just Like a Baby Because Why?

"From the tumor immunology standpoint, the pregnancy model was actually a very interesting one," says Munn. "We chose it because it was kind of an all-or-none readout: the graft survives, or it doesn't. But besides that, it turned out to be helpful because it was so clear to people that the tumor is kind of doing the same thing." Like a baby, a tumor starts out as a little ball of cells that the immune system of the host can't get rid of because the cells are actively preventing their own rejection.

"Just conceptually, the idea that a small thing that is trying to grow bigger can successfully defend itself against a giant immune system that should be able to come in and attack it, and the little thing can do it by turning on the IDO pathway, well, I think that that helped people kind of wrap their head around the idea that you could have mechanisms that create [immune] tolerance starting from a small growing clump of tumor cells."

As for the tie-in to tryptophan, what Munn had initially observed in his co-culture experiment was not a starvation response. Instead, tryptophan was being used as a signaling molecule. This was a sticking point for many of Munn's colleagues. The idea that the amino acid, tryptophan, a component of many types of dietary protein, could serve as a signaling molecule was a conceptual leap that some were unwilling to make.

"Our ability to convince people that tryptophan and IDO was a signal transduction system . . . that didn't really become well accepted until a number of other systems became recognized in the immune system. For example, how a low glucose level is now recognized to have profound effects on T-cell activation," explains Munn, "But when it was first coming out and we were the only example, I think it was easier to assume that we were just crazy than to rewrite the textbooks."

IDO was a tough sell. It was a T-cell suppressor when the concept of T-cell suppression had yet to be established, and it was an enzyme that issued commands by way of tryptophan, an amino acid found in abundance in both turkey and tofu. Nevertheless, sell the idea Munn did, even though it would take another six years to work out the entire signaling pathway that culminated in the suppression of activated T cells. Under this steady stream of data, the paradigms gradually shifted and the skeptics finally bought it. Other labs took up the cause and began experiments that eventually lead to clinical trials with IDO inhibitors, which are now being tested in combination with checkpoint inhibitors, or chemotherapy, or radiation, or permutations thereof.

"That's the way that all of our trials are designed," says Munn, "Pediatric and adult, because the drug is so nontoxic we give the IDO inhibitor throughout chemo and radiation."

As of this writing, more than 20 clinical trials targeting IDO are underway.

✿ ✿ ✿

In early 2015, in a move that suggests a significant vote of confidence in the work of Munn et al., pharmaceutical giant Bristol-Myers Squibb purchased Flexus, a company that specializes in IDO research for upward of $1.2 billion dollars.

DMITRY GABRILOVICH
"MDSCs with T Cell"

CHAPTER TWENTY-THREE

Dmitry Gabrilovich, M.D., Ph.D.

Program Leader, Translational Tumor Immunology Program
The Wistar Institute
Philadelphia, Pennsylvania

MYELOID-DERIVED SUPPRESSOR CELLS (MDSCs)

"Behind the scenes they just whispered, 'Ah, I don't believe it.'"
—D. GABRILOVICH

Dmitry Gabrilovich was born in 1961 in Minsk, then a city in the Soviet Union, and now the capital of the country of Belarus. "My family is there from a long time," says Gabrilovich, his voice still rich with the inflections of his native tongue. "My grandparents from my mother and father's side were both scientists."

Indeed, Gabrilovich's maternal grandfather, Boris Elbert, was a decorated scientist, although somewhat after the fact. He created a vaccine for a disease called tularemia. "He did it actually in prison," says Gabrilovich, rather proudly. "He was sent to prison in 1931 by Stalin, but not to the labor camps, because he was a professor and actually was trained at the Pasteur Institute." Instead, Stalin consigned Dr. Elbert to a special "closed facility" patriotically named The Biotechnical Institute of the People's Commissariat of Defense —a prison for scientists, if you will, where bacteriologists were involuntarily gathered for intensive investigatory comradeship.

> *"[My grandfather] was sent to prison in 1931 by Stalin . . . My mother was born in that place."*

It seems the powers that be needed the vaccine very badly. Before the advent of antibiotics, especially in underdeveloped countries, tularemia (which is transmitted by rodents and their associated insects) had a mortality rate as high as 50%, depending on the strain. "So, he was there for five years, I think," says Gabrilovich, with his dark eyes looking off to the imagined past. "My mother was born in that place."

After a tularemia vaccine was successfully developed, Dr. Elbert was released. A few years later, the Battle of Stalingrad happened. "After the city was liberated it was completely ruined," says Gabrilovich, and that ruination led to rats, and rats led to a severe outbreak of tularemia. But by then, they had a vaccine. "My grandfather, he actually saved the city." Then came the height of irony. In 1948, in recognition of his patriotic efforts, Dr. Boris Yakovlevich Elbert was awarded the Stalin Prize. "That's basically the highest scientific prize in the country," says his grandson. "But he was still officially a convict—by then, a professor in a respected institution—and although his conviction wasn't overturned, he was given that prize."

Political prisoner to prizewinner. Life is funny that way.

Life also has a funny way with love—because Grandpa Elbert, in addition to being a fine scientist was also an amateur matchmaker—at least where Dmitry's father was concerned. "[My father] was actually the best student of my maternal grandfather," Gabrilovich recalls. "So, when my grandfather saw my father, he came home and said to my mom, 'Look, Galina, I found a good husband for you.'" A nice boy, an intelligent young man named Isaak. "And my mother she tells me she said, 'Meh, I have other boys and just don't care about that.' But then they met, and, lo and behold, got married."

Off to School, and HIV

Dmitry Gabrilovich's initial training was completed in 1984 at Kabardino-Balkarian State University, Nalchik, USSR, where he specialized in infectious diseases. Afterward, he moved on to a fellowship at the Central Institute of Epidemiology in Moscow in the mid-1980s. "I did my clinical training in Moscow and continued research there," says Gabrilovich, "and this is when the HIV epidemic starts."

Gabrilovich was able to secure funding from the government and set up a small, but well-equipped research facility for the study of HIV, which back then was a bit of a tricky thing. "At that time we didn't have AIDS in the Soviet Union," Gabrilovich explains, at least

> "At the time we did not have HIV in the Soviet Union [it was not permitted]."

not "the gay kind." Transmission of HIV via blood transfusion was "allowed," but homosexuality did not officially exist. That was the story.

In 1988 Gabrilovich saw his first HIV patient in the clinic, a patient well within the Soviet messaging of the day. "He was from Africa, and happened to be in Moscow at the time." But the official line quickly became a border to absurdity. "Then of course we found a male prostitute," says Gabrilovich, with a bit of a finger wag. "He was an officer in the Soviet army and he infected hundreds of people." That's how it started. "And now it's become a huge problem for Russia, as we all know."

While in Moscow, Gabrilovich made research headway and published a few papers regarding the prognostic value of **neutrophils** in HIV. Despite this, and contrary to the obvious threat of a rapidly expanding epidemic, continued funding was hard to come by. Gabrilovich was forced to move on.

Neutrophils are the most abundant type of white blood cell in the immune system, and are one of four cell types classified as granulocytes. Neutrophils kill foreign invaders by both phagocytosis (a process whereby the invader is engulfed and dissolved) and degranulation, whereby neutrophils park next to an invader and release cell-killing toxins. To have too few neutrophils is to be "neutropenic," a condition often brought on by chemotherapy that leaves patients highly susceptible to infection. Neutrophils are indeed alerted to the presence of HIV-infected cells, but they are unable to clear the infection.

Aha! First Contact with MDSCs

As with many of the narratives in this book, a mentor stepped in to take the story forward. "Stella Knight—I wrote her and say that I want to study dendritic cells—and she graciously agreed to help me write an application to Wellcome Trust," a United Kingdom–based charity that provides grants for biomedical research. The current endowment, established by pharmaceutical magnate, Sir Henry Wellcome, exceeds £20 billion.

As his mentor would have it, Gabrilovich got the grant, and put it to use by taking his rapidly expanding skill set to London where he worked on HIV's effect on dendritic cells (DCs; see Steinman, Chapter 10) and then on to the University of Texas Southwestern and the lab of David Carbone where he explored DCs in cancer. It was during this work that myeloid-derived suppressor cells (MDSCs) were first encountered.

It was a semi-aha! moment, recalls Gabrilovich, caused by the unexpected results of an experiment. "In 1995–96 we demonstrated dendritic cells are defective in cancer, so we published a big paper" (Gabrilovich et al., *Cell Immunol* 170: 101, 111 [1996]). A logical follow-up to this work was to explore DCs in the presence of VEGF, and that's when things got really interesting.

Vascular endothelial growth factor (VEGF) is a protein: a cytokine signal released by certain cell types that causes new blood vessels to form. VEGF is activated during embryonic development, after an injury, or following vigorous exercise. VEGF signaling is also used by tumors to demand new blood supplies for their continued growth, a phenomenon that has been the focus of numerous drug development programs. One such program produced bevacizumab, a blockbuster, multi-billion-dollar drug currently being tested in combination with checkpoint inhibitors.

"I put VEGF into the mouse in large quantities just to see what happens with dendritic cells," Gabrilovich explains. "So, we find that dendritic cells were not functional in the presence of VEGF. That was what we expected, but what was also happening is a large expansion of cells which had for me, at that time, an unusual phenotype."

Cell types in biology are identified by phenotype: how they look, how they act, and how they affect their surroundings. Gabrilovich was looking at something he'd never seen before. He turned to the literature for guidance and found references to Gr-1$^+$CD11b$^+$ cells, indicating that the cells belonged to a myeloid lineage (i.e., originating in the bone marrow, as many cells types of the immune system do). However, the cells he observed did not phenotypically belong to any of the mature myeloid-derived cell types, like neutrophils. "I was quite curious about this," says Gabrilovich, "because it had never been described before in a VEGF setting." Further reading revealed previous descriptions of similar cells by Hans Schreiber's group at the University of Chicago and Rita Young's work at the Medical University of South Carolina: cells that were more or less the same, but not really, suggesting to Gabrilovich he'd seen something new.

"So, I work on this in '97, and in '98 we published a paper in the journal, *Blood* [Gabrilovich et al., *Blood* 92: 4150 [1998]]. And then a year later Vincenzo Bronte [of Verona University], he published his paper in the *Journal of Immunology* [Bronte et al., *J Immunol* 161: 5313 [1998]] showing cells with exactly the same phenotype." So, there's this semi-new, semi-different (but not exactly) cell type. This distinction is important because if you remove them, says Gabrilovich in subtle crescendo, index finger in the air, "you dramatically improve vaccine response." These cells, whatever they were, were immunosuppressive.

To Name Is to Know

Shortly thereafter, Gabrilovich established his first independent lab at Loyola University in Chicago, but soon moved to the Moffitt Cancer Center in Florida. Both Gabrilovich and Bronte continued to expand on their initial novel observations, but a problem soon became evident: The nascent field was getting out of hand. There was lots of activity but no discipline. There was too much information. "Everybody came up with their own descriptions, their own explanations," says Gabrilovich, "And consciously I tried to fix that."

Gabrilovich e-mailed the entire universe of experts (all seven of them) in the as-yet-unnamed cell type and insisted that the subject of their investigations have a name. "And I don't care what name it will be, but we need to have name because otherwise we will be buried with so much stuff."

Gabrilovich reached out in particular to Suzanne Ostrand-Rosenberg at Maryland University, who came up with the name "myeloid-derived suppressor cells," which Gabrilovich was then compelled to run past the English Department at the University of South Florida. "I called the linguistics department because there was a debate whether we could call them 'derived.' What does this mean, the use of the word 'derived' in proper science?"

Why so precise? "Because Hans Schreiber was very picky on the names," says Gabrilovich. Granted, it wasn't the most rigorous exercise in nomenclature, given all the new data that started to pour in regarding the cell's general characteristics, "but we all agreed that it's probably the best at that moment. So that's how we named the cells." Shortly thereafter, the journal, *Cancer Research* published the new name in a 2007 commentary (Gabrilovich et al., *Cancer Res* 67: 425 [2007]).

Can You Hear Me Now?

After the commentary, the field took off. The publication rate of MDSC-related research papers went from 10 to 12 a year to 10 to 12 a month, and then per week. "And they've just been absolutely nonstop since 2008," marvels Gabrilovich. "First in cancer, then for lots of infectious diseases. HIV now has become quite populated with the stories about MDSCs. The rest is history."

> "First in cancer, then for lots of infectious diseases. HIV now has become quite populated with the stories about MDSCs. The rest is history."

Which is not to say the victory was easily won. "In 1999, 2000, we were alone," says Gabrilovich. "Me and Vincenzo Bronte." Their observations and related implications escaped even the most highly respected, capable researchers, even Hans Schreiber, who had actually shown the cells' existence in a paper in the mid-1980s but hadn't recognized their importance. "He didn't believe it at that time. He actually even told me, 'My goodness, I didn't appreciate it. If I knew, I would have focused on that more.'"

Others at the time simply dismissed MDSC literature as confused observations of what were actually immature **macrophages**. "I mean, they couldn't argue in front of me, but behind the scene they just whispered, 'Ah—I don't believe it.'"

A __macrophage,__ derived and differentiated from a myeloid cell progenitor, is a component of the immune system that has no particular target in mind. Macrophages detect, engulf, digest, and destroy all sorts of materials that are generally classified by the immune system as foreign. This includes microbes, cellular debris, cancer cells, and foreign substances. Macrophages differ from dendritic cells in the sense that the macrophage is a beat cop, and a dendritic cell is more like a detective—more focused and more authoritative when it comes to arresting a suspect. (Apologies here to science mavens—the differences between the two cell types are as subtle as they are complex. The metaphor saves 1000 words.)

"You cannot imagine how many problems I had personally," says Gabrilovich. Even now, hundreds (thousands?) of papers later, there are still those in the field that can't accept that a critical aspect of their life's work had eluded them. "We still have detractors from the entrenched old generation, especially the macrophage field. People who worked on macrophages their entire life, they couldn't accept that there is a huge wave of immature myeloid cells which belong to the same lineage." MDSCs continued to be dismissed as a transitory state, indistinct: teenagers that go on to become macrophage adults.

But Gabrilovich kept pressing, and the data were not to be further denied. "MDSCs are not existing in normal individuals, only in pathological conditions," stresses Gabrilovich. "They are critically important in chronic inflammations, like autoimmune diseases or cancer." The biotech industry agrees. At the time of this writing, there are more than 20 clinical trials in cancer underway with the detection or targeting of MDSCs as a component of the investigational protocol. Furthermore, the MDSC target could be low-hanging fruit in terms of drug development because there are already approved agents that have anti-MDSC activity. One is gemcitabine, a chemotherapy agent that, as a cytotoxin, seems to work against MDSCs. Another drug that has activity in MDSCs is the erectile dysfunction drug, sildenafil.

Suffice it to say that excitement regarding MDSC-related immunotherapy is real, and it continues to rise.

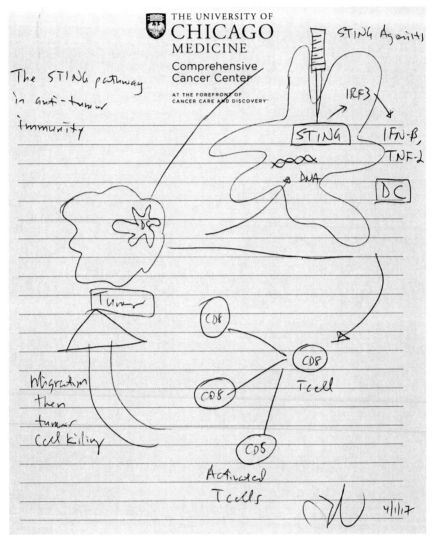

THE UNIVERSITY OF
CHICAGO
MEDICINE
Comprehensive
Cancer Center
AT THE FOREFRONT OF
CANCER CARE AND DISCOVERY

The STING pathway in anti-tumor immunity

STING Agonists

IRF3

STING

IFN-β, TNF-α

DNA

DC

Tumor

CD8

CD8

CD8

T cell

CD8

Migration then tumor cell killing

Activated T cells

4/1/17

TOM GAJEWSKI
"STING"

CHAPTER TWENTY-FOUR

Tom Gajewski, M.D., Ph.D.

Professor of Pathology and Medicine
University of Chicago
Chicago, Illinois

STIMULATOR OF INTERFERON GENES (STING)

"We're not engineering anything. We're trying to recapitulate how Nature does it successfully when Nature does succeed. It's sort of an Asian approach. Like Zen." —T. GAJEWSKI

Tom Gajewski is 55. He was born and raised and educated and trained and works and lives and loves his family in Sweet Home Chicago. "There are a lot of shootings these days out in the rough neighborhoods . . . that's a little bit disconcerting," notes Gajewski. "It is a great city, though."

Chicago really is amazing. There's the magnificent architecture, the museums, the river, the lake, restaurants, ballparks, clubs, and then those special, personal places. "We live in a neighborhood right near the University of Chicago," where he and his wife, Marisa Alegre, M.D., Ph.D., have research labs. Together they have a house and a garden with a deck for cookouts when the weather is right. "A place to sip a glass of wine with the smell of the flowers as the sun is setting right in the middle of this urban environment." It's his refuge.

"You'd be surprised how many oncologists garden," says Gajewski. They do it almost out of necessity. To come home and tend to something vivid, something thriving, something so very much alive is an essential counterpoint to the days in the clinic.

It's unusual for a scientist of Gajewski's caliber to have completed nearly his entire education and clinical training in the same locale, yet, in addition to the many cultural lures of Chicago there were two very powerful sensory attractants that explain Gajewski's lack of wanderlust. "When I started looking at different universities, I'd never actually been to the University of Chicago," Gajewski recalls, "but when I came here it was immediate, I just had a feeling about the place." That feeling was generated by the physical aspects—the neo-Gothic architecture that instilled a sense of timelessness—as well as the denizens of those weighty structures that Gajewski met as he interviewed. They gave the impression that this was a place where knowledge was accrued for the sake of knowledge. For the scientist-to-be, it felt like a shrine for people who crave a good think.

That was almost enough for Gajewski, but what really convinced him to stay was this: "So, this is 1980 and I had never heard live blues music in my life," and Chicago is pretty much ground zero for that sort of thing. "The first week of undergrad we had an orientation party in our dorm here on campus, and the band was—get this—*Buddy Guy*," says Gajewski, eyes wide. *The* Buddy Guy, blues legend. "The guy was great, so energetic . . . and now of course, he's famous. In fact, this past year I was at a meeting in Montreux, Switzerland, which is where an annual jazz festival is held, and he's one of the recognized heroes, locally." Gajewski smiles, marveling at the memory. "This guy played at my dorm party." That guy. Buddy Guy.

> *'The first week of undergrad we had an orientation party in our dorm here on campus, and the band was—get this—*
> ***Buddy Guy"***

Gajewski's interest in music is not merely as an enthusiast; he's also a practitioner. He started playing the guitar at the age of eight, and he's currently in a rhythm and blues band, and yes, he's pretty damn good: he's the lead guitarist for The Checkpoints (see Chapters 1 and 15).

Mentor at the Center

Architecture and Buddy Guy aside, as with many of these stories, there was a mentor. For Gajewski, it was, unquestionably, Frank Fitch. "He's the founding father of immunology on this campus. He made many of the fundamental discoveries of how T cells work." What most struck Gajewski, however, was that Fitch did not simply impart knowledge as scholars often do, he had the grace to elicit it.

Gajewski encountered Fitch in his common core biology class. "This very established guy was teaching a very basic introductory-level science class," called Defense Mechanisms. The course had an immunology focus, and the syllabus covered a lot of information, but passing the class had nothing to do with rote learning; Fitch was Socratic. "He allowed students to learn on their own, to discover the answers in the middle of class," explains

Gajewski. The motive was not to achieve learning utterly de novo—one still needed to prepare for class with background reading—but Fitch was not there to goad his students. He was their guide.

"In the beginning of the quarter he posed some questions, and the result might be five minutes of total silence," remembers Gajewski. These were long, painful minutes as Fitch patiently waited. "It was pretty clear he wasn't going to answer the questions for us, so once everybody got used to that, then we started to talk." It was a classical method of self-discovery. "I'd never had a class quite like that. So, yeah, shortly after that I started working in his lab," completing both his undergrad and graduate work with Fitch.

Starting Off on the IO Road

"What he was doing at the time would never be NIH [National Institutes of Health]-funded now," notes Gajewski. "Especially with the payline being so low, there's much less funding of high-risk projects." For Fitch, the risky project was trying to figure out how to clone T cells and grow them in tissue culture, something that had never been done before. Once done, the clones would be an invaluable resource for probing T-cell type and function. (At least eight types of T cells have been identified, differentiated by their varying functionalities.) "Some T cells would proliferate when they saw antigens. Some would kill their target when they saw antigens," says Gajewski, "And they figured this out in the 1970s." Given the technology available at the time, this is like saying you found a new planet by looking at the sky.

The risk of failure for such a project was high; even the tools to address the question had to be created. Fitch turned to the then-new technology of monoclonal antibodies, generating his new constructs by immunizing rats with mouse T cells. The rat's immune system, on seeing foreign mouse T cells, will design antibodies to do away with them. The assumption was that when you immunize with whole cells, which no one was really doing at the time, you get antibodies that recognize cell surface structures: the structures that interact with target antigens.

The project was a roaring success. "They generated antibodies against CD4, CD8, adhesion molecules like LFA-1, CD45, and the T-cell receptor," says Gajewski, all of which are critically important cell surface proteins on T cells. "Many of the antibodies that we buy now from companies to do flow cytometry [a method of sorting cells by type] are antibodies that he made. So those were the early days of characterizing T-cell structures," and Fitch was the expert. In 1995, the Frank W. Fitch Monoclonal Antibody Facility was created at the University of Chicago to promote this technology.

Doctor, Doctor

Gajewski's graduate work with Fitch, and the associated publications documenting their successes, was attention-getting. "I had invited talks as a graduate student . . . I gave a talk at Berkeley," invited there by James Allison, no less (see Chapter 1).

After graduate school, it was back to the scholastic grindstone. Gajewski was on the M.D./Ph.D. track, and he still had the M.D. to complete. "You're demoted when you go back as a medical student," Gajewski recalls, "Because now you're not special anymore. You're just one of the 100 people in the medical school class." But graduate work gave him an edge; he was a few years older, and therefore more mature than his classmates, and he was experienced in research and at ease with the unknown.

"The medical students were just trying to get through it so they could practice medicine," recalls Gajewski, "But the research-oriented medical students were looking for questions. That's a little bit of a different viewfinder." With that mind-set, Gajewski approached his medical school rotations. "Every rotation, I was looking for questions. My psychiatry rotation I was like, 'Wow, psychoneuroimmunology! That could be really interesting . . . the mind–body connection. I could work on that!'" Or there was the surgery rotation where he was mentored by a famous thyroid surgeon. Thyroid cancer and the immune system? That's a thought. Or diabetes, or infectious disease, or rheumatology. "Every rotation, I found some immunologic application."

A career was coming into focus. "What happens when you do all your clinical rotations is you realize what fits your lifestyle and what doesn't," says Gajewski. "I was pretty certain I didn't want to be a surgeon, didn't want to do OB/GYN, didn't want to do anesthesia, didn't want to do radiology." As he whittled down the choices the totem of internal medicine emerged and, finally, a rotation in oncology. "I was speaking with Everett Vokes, one of my attendings here [now the Chair] and he's like, 'Well, cancer is the hardest problem. Why wouldn't you focus on cancer?'" That was the hook. Gajewski was looking for problems to solve, so why not tilt at the tallest windmill? "So that all clicked and came together."

For the Graduate Student Who Has Not Slept in Days

Before discussing Gajewski's work on the STING (stimulator of interferon genes) pathway, the work for which he is best known, there is something that first must be said:

This is Your Life.

"This is how we advise students now. You have to enjoy it while you're in it." Your education, your training, it's not some obstacle course to be endured. "It's not just a rite of passage like a hazing or something," says Gajewski. If that's the way you see it, you're doing it wrong. "Look, I got my first job when I was 35, but I knew going in that it was going to be that way." An M.D./Ph.D. program entails more than a decade of training. A decade should not be, *cannot* be treated as a protracted boot camp. A decade is, if you're lucky, an eighth of your stay on this planet.

"*This is your life.*"

"I think most students actually can't zoom out and look at the big picture," says Gajewski, "But I was in no hurry. I was enjoying each step . . . research in the lab, making discoveries. It was just fun." If you don't think it's fun, if you can't see the glory of the forest for what you think are menacing, tangled trees, then you need to get out of the business. "Occasionally you'll have a student, and it will come out in conversation, this sentiment that, 'Oh, I've just got to get through this.' And I'm like, 'No, no, no, no, no. Someday you're going to look back with fondness at your student years.'" If you've made the right choice, if you're where you're supposed to be, then embrace that circumstance. Revel in it.

"This is your life."

In the Beginning

His M.D./Ph.D. program completed, Gajewski's next step was to train in tumor immunology. Although this did not require that he leave Chicago, there were other forces in play that prompted a move. Gajewski had fallen in love. "I met Marisa, who is now my wife, during my Hematology/Oncology fellowship," says Gajewski. At the time, she was doing research in Jeff Bluestone's lab (University of Chicago; see Chapter 21). "We met and then she stayed for a Ph.D. while I was finishing up my medicine training."

Marisa was from Brussels. The couple decided to give the city a try as a potential home base. "I knew that I wanted to do cancer, and that I wanted to go to Brussels." He consulted his mentor, Frank Fitch, for suggestions as to whom he might work with overseas. Fitch named Thierry Boon, a prominent cancer immunologist. Gajewski was accepted into Boon's lab and spent the next two years working in Belgium. "That's where I learned real tumor immunology and tumor antigens," says Gajewski. "He's the father of molecular identification of tumor antigens."

Gajewski spent two years under the distinguished wing of Thierry Boon and then returned to Chicago fiercely prepared to cure cancer. "That's when everything had come together with identified antigens and a new strategy to vaccinate," says Gajewski. "I hit the ground running."

The initial plan was to add an adjuvant (a substance that boosts a vaccine's efficacy), a cytokine called IL-12, into the vaccine mix to ramp up T-cell activation. The approach resulted from the work in Boon's lab: They had taken T cells from a cancer patient and grown them in a dish; they then took tumor cells from the same patient and grew those up in another dish. When the two cell types from the same individual were then combined in the same dish, the T cells killed off all the tumor cells. "Both cell types came from that patient where the tumor is growing happily," yet Gajewski had just observed that same patient's T cells massacre tumor cells in his lab. This generated a crucial question: "Where's the block in the patient?"

One potential block was that, in the setting of cancer, T cells might not receive all the logistical support they needed, thus, the addition of IL-12, a cytokine produced by dendritic cells that turns inactive T cells into T cells ready for battle. Gajewski tested this idea clinically on patients with advanced melanoma using a vaccine comprised of known melanoma antigens combined with IL-12. "We found we could induce T-cell responses in the blood, and they were quite potent." That was the good news. The bad news was that only a few patients experienced shrinkage of their tumors. "The results pushed the question to the next level: What could be downstream from failed induction of T-cell response?"

The question was timely. The technology to explore the question had just come online. Small chips—so-called "microarrays," the size of a business card or so—had just been developed that could tell you which genes in a cell were active. Activated genes can tell you what cell type you're dealing with and what the cells are up to. Gajewski's microarray analysis showed that many patients have a preexisting immune response against their tumor, meaning that the genes that were turned on were indicative of T-cell activity. "So there already was an immune response there," Gajewski explains. "That was about a third of the patients, and when we did get a clinical response with the vaccine it was in that subset, those patients who had tumors that allowed T cells in. For us, this is really a breakthrough moment." The epiphany implied that immune interventions (e.g., vaccines) preferentially occur in patients who already have an ongoing immune response. Further, it suggested that primary resistance to immunotherapies occurred in the tumor microenvironment in which the dialogue among activators of the immune response was somehow interrupted.

From these observations, Gajewski formulated four questions, four avenues of investigation that have kept him busy for the last ten years.

- How does an immune response, effective or otherwise, occur without the immune system sensing the presence of a pathogen? (Tumors are not pathogens. Tumors are "you.")

- If T cells are present in the tumor (which is not a pathogen), what brought them there?

- If T cells have managed to somehow detect the tumor and move in, why are they so often ineffective?

- How do some tumors exclude T cells altogether?

Gajewski came to STING by addressing the first question, which can be rephrased thusly: How does the presence of a tumor sound the alarm of the body's immune response? One hint was that Gajewski had seen type I interferon signaling (indicating an immune response) in the few cancer patients that had had a clinical response to the vaccine. This signaling is indicative of an innate **immune response**, as opposed to an adaptive immune response.

Expanding on the definition in Chapter 7, the innate immune response is triggered by components of the immune system that are able to identify potential threats based on generalized criteria, referred to as DAMPs and PAMPs (damage-associated molecular patterns and pathogen-associated molecular patterns, respectively). PAMPs and DAMPs are evidence that you've been invaded. There are any number of receptor molecules in the innate immune system attuned to one or more of these patterns — these signals — that something in the cell is amiss. For example, toll-like receptors (TLRs) come in at least nine different types, and each type recognizes a particular PAMP or DAMP. In addition to the TLR system, we have STING, a receptor capable of recognizing cytosolic DNA — where DNA should not be.

Oh Death, Where Is Thy STING?

The detectors used by the innate immune system to spot infectious disease are well known. Many of these are referred to as a class of molecules called toll-like receptors (TLRs). TLRs sense the bits and pieces of which microbes and viruses are made, including RNA and DNA. Different TLRs detect different bits and are present in a variety of locales, be it a cellular compartment or on the cell surface. When a TLR detects an invader, it sends out a type I signal, which brings in antigen-presenting cells from elsewhere in the body to survey the breach, gather information about the invader, and then educate T cells on what to track down and kill.

Taking advantage of this wealth of knowledge and investigative tools, Gajewski proceeded to create mutant cancerous mice for every known invader-sensing entity. "All we had to do then was line up the knockout mice for these pathways and see which knockout lost the spontaneous T-cell response against the cancer." The only mouse that lost the response had a defect in the STING pathway. STING, it turns out, detects aberrant DNA that has found its way to the cell's cytoplasm. (To be absolutely clear, STING was not discovered by Gajewski. That honor goes to Glen Barber at the University of Miami. Gajewski did, however, leverage STING in the fight against cancer.)

"It's not a TLR, and it's probably just an adapter," Gajewski explains. The direct sensor for cytosolic DNA identified so far is called cGAS (cyclic GMP-AMP synthase), which—when it senses alien cytosolic DNA, as would be the case in a cell infected with bacteria or virus—generates a second messenger called cGAMP. This turns on STING, which passes the message to IRF3 (interferon regulatory transcription factor 3), which in turn orders the cell to churn out type I interferons. "We found that natural immune responses against tumors were lost in the STING knockout mice."

So, STING is one of the ways Nature does it. This, to Gajewski, begged the next question, "If we developed a drug to push that pathway, could we over-drive an immune response or initiate one where one didn't exist before?" The potential answer to that question is now the subject of investment by the biotech industry, specifically, Merck, and a company called Aduro. "They were developing STING agonists as a vaccine adjuvant," says Gajewski. "I showed them our data and we realized STING agonists could be a direct cancer drug." Working in collaboration with Aduro, several STING agonists were identified and are now undergoing clinical testing. "When we tested them in our animal models, they had the most potent antitumor activity we've seen with any immunotherapy."

> "I showed them our data and we realized STING agonists could be a direct cancer drug."

Case Report

The drug referred to above is the subject of a Phase I clinical trial in melanoma patients that commenced in May 2016. Phase I investigations typically involve what's called dose-ranging, from low to high, as investigators try to gauge the proper dose for safety and efficacy in humans. "It's just the first dose level," reports Gajewski, which is the lowest, but he's already had a very encouraging glimpse at the drug's potential efficacy in one of his recently treated advanced melanoma patients. "She's already failed ipilimumab and anti-PD-1, so that's the patient population we want to try and get a new immune response going in," Gajewski explains. "When we looked at her tumor we don't see any T cells in it, so this is the highest bar. This is the leftover population where you've got nothing going endogenously and we've got to try and trigger something new."

The patient had palpable lesions on her leg. According to study protocol, the STING drug was injected directly into those tumors. "She just called and said she had a fever and the nodules on her leg were all red and warm," Gajewski reports, excitedly. "I asked her to take a picture, right? Because who knows . . . she comes to clinic again next Tuesday, it might be gone. So, on my phone right now I have pictures she sent and the injected lesions are all red, and the *noninjected* lesions are all red," which is exactly how a vaccine like this needs to work. You cannot treat every tumor in patients with advanced disease: the tumors are too many, too small, too inaccessible. Red means an inflammatory immune response was raging. It was a thrill.

"It's very early," Gajewski freely admits. "Who knows if it's going to work, but it's just the way we think about it: How did Nature set up the natural immune response? We're not engineering anything. We're trying to recapitulate how Nature does it successfully when Nature does succeed. It's sort of an Asian approach. Like Zen."

Music Man

Enough science. Let's rock.

Gibson, Fender, or Les Paul?

"I've been a Les Paul guitar guy for versatility. It's a heavier guitar, it's more solid wood and so it can be a richer sound because of that. Another reason is because I have this sort of magical Les Paul with a perfect action. I haven't found another guitar that has that perfect action for my feel, for my fingers. I have to say, though, that lately I've been looking at a few other guitars. I did get an autographed Fender Stratocaster from a fundraiser recently, which includes Sting (*the* Sting) as one of the signers."

Sting for STING. How cool is that?

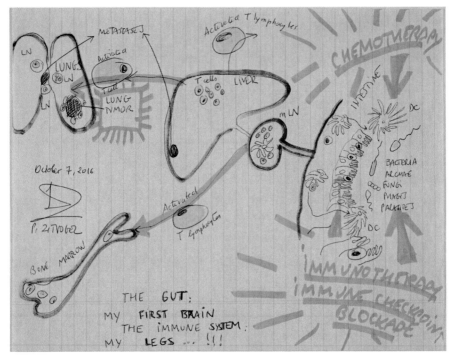

LAURENCE ZITVOGEL
"My First Brain"

CHAPTER TWENTY-FIVE

Laurence Zitvogel, M.D., Ph.D.

Research Director, Tumour Immunology
and Immunotherapy
Gustave Roussy Cancer Center
Villejuif, France

MICROBIOME

"You can be an outstanding scientist if you have a lot of, you know, des grains de sable dans les rouages." —L. Zitvogel

Note: Should there be any gender confusion caused by the first name, Laurence Zitvogel is a woman, and according to many who were interviewed for this book, she is also "very French."

Laurence Zitvogel was born in the commune (township) of Suresnes, a western suburb of Paris, in 1963. "It was an unfavorable place, because it was communist," recalls Zitvogel of those days when the dark blue work jackets favored by party members were a common sight. There was a history of political unrest in the area, as well as nearby in Nanterre; the seeds of the 1968 student riots were sown there.

Zitvogel's parents were artisans and painters. Living was modest, but there was love and life was good. Others in her circle were not so blessed, particularly in Nanterre, an area with a large immigrant population. "I was very privileged because I could go on vacation with my parents," says Zitvogel, "But they were not going anywhere," and for the least fortunate among them there could be no escape. "One of my friends was very gifted. He was from Algeria originally. His brother was shot to death by a white man."

There was tension to be sure, political and racial, but there was also an education to be had, a core value to be crafted. "There were these two populations, the whites and the Algerians, and then the communists in the background, a mixture that created a very hot social atmosphere," says

Zitvogel. "It was very good for the mind. I was encountering people with a lot of courage, a lot of willingness to become somebody in life." It wasn't a matter of fairness; it was a matter of the fight. "It was a very good place for understanding that your life is in your hands—and in your genes also—but if you do not have luck at the beginning, life might be more difficult for you." You might have to be bold. Your situation might someday become utterly unacceptable, as it eventually did for Zitvogel. Then, you have to act decisively.

Doctor: No-Brainer

There wasn't a second thought. "I was lucky enough to be born with a vocation. By the age of six or seven I wanted to be physician." There was no family precedent for it: artisan parents, uncles in the trades. Grandma was a shoemaker. "Nobody was a physician. Nobody was a biologist." Why then choose to be a doctor? Zitvogel doesn't know. "This is something that I never could explain. I just wanted to be by myself, be independent, and be capable of saving life. I was determined. I don't want to be here [in Suresnes]. I don't want to die without palpable achievement. It is impossible to consider. I could not live just for myself—for maintaining or entertaining myself. This is not something that I like to think about. Each step I am doing in my life it is to achieve a goal, and the goal is to progress in medicine, and to progress in medicine you need to achieve scientific progress because still we don't understand what we are doing."

She set off to Paris to become a doctor and, in short order, a doctor very much immersed in the phenomena of the immune response. "I was fascinated by vaccination against infectious diseases, so vaccination against cancer was something I thought about very early on," even before she had any formal training. "I wanted to vaccinate against, you know, transformed cells. I think I might have read that somewhere. [Quite possibly in Steve Rosenberg's book, *The Transformed Cell*.] I was fascinated by immunology as soon as I had my first course: antibodies, B cells, and T cells."

There might also have been a bit of national pride in the attraction. Awarded during her early training, the Nobel Prize in Physiology or Medicine in 1980 was bestowed on three individuals: Baruj Benacerraf (Venezuelan by birth), George Snell (American), and Jean Dausset (a Frenchman). The award recognized the trio's "discoveries concerning genetically determined structures on the cell surface that regulate immunological reactions." Meaning, they got the award for figuring out the thing that tells T cells what to kill. It was a critical breakthrough in understanding just how the immune system does its job.

Eyes Opened to IO

Zitvogel's first choice of specialty involved a rotation in internal medicine, which also contributed to her attraction to immunology. "This was crucial because we learned how difficult it was to handle lupus erythematosus," a potentially fatal autoimmune disorder. And there were related lessons about other autoimmune diseases.

Then came a rotation in cancer: hematology—lymphomas and leukemias—the cancers of the immune system where B cells and T cells go awry. "When I came to know about hematology, I discovered that this is what I need to do to dig myself into immunology," says Zitvogel, who at the time was greatly intrigued, fervently hopeful, and deeply determined to make her mark. Once in the clinic however, her enthusiasm quickly hit a wall.

Zitvogel's introduction to the therapeutic options for treating cancer patients was nothing short of offensive. "These oncologists would just deliver this stupid medicine, this chemotherapy into patients, this *poison*, and they would not even understand what was mechanistically going on." To Zitvogel, it could barely be called science. "They were only pharmacologists, you know, measuring the levels of the secondary metabolite of this or that pro-drug." It was absurd, and the side effects were devastating. "The patients were suffering. I could not bear this ignorance, this stupid mentality." There was no scientific rationale for choosing regimen A over regimen B. You might as well write the drug names on a handful of dice and give a throw. "That is what I discovered. It was a miserable, miserable education in oncology. I was so pissed off by the whole thing."

"It was a miserable, miserable education in oncology. I was so pissed off by the whole thing."

By the mid-1980s, the research activities of Steve Rosenberg at the National Cancer Institute were the stuff of headlines. The immune system stimulant, interleukin 2 (IL-2), was the latest magic bullet. "I heard about Steve Rosenberg with the IL-2, and the tumor-infiltrating lymphocytes and how you can reinvigorate the immune system against cancer and I said, 'Ah! That's a concept. That's a vision. I believe in it.'" Yet, tumor immunology as a discipline did not really exist at the time. The science was in its infancy, and what was known (which wasn't much) was impossibly complex.

Nevertheless, Zitvogel was sold. "I decided to interrupt my medical education because I needed to get myself into tumor immunology, and I said, well, if it takes a lifetime to learn, it takes a lifetime. I was 25 when I decided to engage in that." Zitvogel began training in Paris with Wolf-Herman Fridman at INSERM, one of the pioneers of IL-2 and LAK (lymphokine-activated killer) cell investigations. The association only lasted a year. "Our relationship was not very good. I had my own ideas and would not necessarily agree with his suggestions. It was tough."

With a contentious relationship and a failing project, Zitvogel decided to get closer to the source, closer to Rosenberg. For this she signed on with Mike Lotze at the University of Pittsburgh, who was, as Zitvogel puts it, a metastasis from the Rosenberg lab, part of the diaspora of talent that flowed (still flows) out from that special place at the NIH. "He's a very nice man, Mike, very enthusiastic, tons of ideas," says Zitvogel. "He wanted to do everything, which fit in nice with my view because I wanted to do everything to broaden my knowledge, to be open to all the possibilities of how to arm the immune system against cancer." The "everything" they did included work on the cytokine IL-4 and leukemia, therapeutic ideas for dendritic cells, gene therapy with retroviral vectors, IL-12, the B7.1 signal—you name it. The possibilities seemed endless, and it was full speed ahead.

Unfortunately, some people had a problem with that. The problem was not a matter of methodology, but rather with the underlying philosophy of Lotze's mentor, Dr. Rosenberg, who was also experimenting with every immunologic thing under the sun. Everyone agreed that Rosenberg's work was cutting-edge, but practitioners of cutting-edge technology that push new ideas as quickly as possible into the clinic (perhaps too quickly) are called "cowboys." And some cowboys are called reckless.

Zitvogel sees it differently. "Steve is a surgeon and, presumably, just like me, he hates to see his patients dying." So, despite the odds, you do everything you can. You try. If you fail, you try again. It's the darkest sort of frustration but you keep going. "It's when you cannot do anything to help somebody, and this person is young or has a family or has this life that has been torn apart, and you are supposed to have the knowledge," says Zitvogel. "You're standing there in your white clothes in front of somebody who is in bed, naked, suffering, dying."

Without doubt, there is still a massive amount to learn about cancer immunotherapy so that it can be both safe and broadly effective, but until that potential comes to pass you still have to do *something*. You put on your cowboy hat. You saddle up. You try.

Moving On, Moving In, Bugging Out

"Dendritic cells (DCs; see Steinman, Chapter 10), that was my second love," and a love directly related to Zitvogel's interest in vaccination. The work began with Mike Lotze and lasted for five very productive years. On Zitvogel's return to France, investigations continued with her new partner, both in science and in life, Guido Kroemer. This was the mid-1990s and Kroemer was already a highly regarded researcher in France. "I met Guido around that time. He was the king of **apoptosis** and apoptotic mechanisms."

Apoptosis is the process of programed cell death. Generally speaking, it's like when leaves fall in autumn. Winter is coming and the tree doesn't need leaves anymore so it initiates a genetic program leading to leaf loss; this same process is also used to get rid of diseased leaves or leaves that are damaged: The tree deliberately kills them off and they fall. In animals, the analogous process is called apoptosis. The body has determined that, for whatever reason, the cell should die. The body instructs the cell to do so and it does, although not all at once; the cells don't pop, but rather disintegrate in an orderly fashion. Some chemotherapy drugs can also induce this tidy demise in a manner that is immunogenic, meaning that the immune system can screen the by-products of cellular disintegration. This is exactly the way you want cancer cells to die: to undergo immunogenic apoptosis.

The manner of cell death is critical to getting DCs to effectively display cancer antigens, so Kroemer's work was a natural continuation of Zitvogel's research with Lotze. The goal was to reveal the links between a certain type of cell death—immunogenic cell death—and the effective processing of that cellular corpse by DCs that turn the molecular information into seek-and-destroy orders for killer T cells.

The work went on for 12 years. Discoveries flowed. "We discovered that different chemotherapies mediate different qualities of cell death," recounts Zitvogel, "We discovered the concept of immunogenic cell death and the molecular and metabolic cues underlying this concept." The discoveries were transformative in the field, and the scientific power couple became renowned for their achievements.

However, once gained, that hard-earned recognition became a significant liability. "I applied to a very competitive European grant," explains Zitvogel. She really needed the money for her research. She didn't get it. "I was accused of not being independent because I spent the last 12 years co-working with Guido, my husband, even though I was independent on all the other things I was doing." And those other things were also critically important to the field: DCs, NK (natural killer) cells, and research into what is now the incredibly hot topic of something called **exosomes**.

Exosomes are vesicles: tiny bags of cellular material that periodically bud off from the surface of both healthy and cancerous cells. These bags—more like bubbles, really—are formed from the outer membrane of cells and are filled with materials from inside the cell, like metabolites, signaling molecules, RNA, DNA, and cellular debris. Once packaged, these exosomes are released into the bloodstream where, for a time, they circulate. The information represented by the material inside these bubbles is now the subject of intense investigation, as they are a potential treasure trove of information about cells—in particular, cancer cells—that could possibly be used to construct a vaccine.

"My grant was rejected twice in a row on the basis that I was not autonomous," says Zitvogel, and further, she feels that gender played a role. "When you are a female and you are bright, you have to be dependent on the male," that's the way it works, or at least, that's the way the world often sees it. "Whereas actually, it is the other way around; often males are dependent on an important female that is not visible." The grant reviewers dismissed Zitvogel's application because they could not see her. It was an outrage.

"I talked to Guido. I said, 'Look, this is incredible. This is our society. It's completely crazy. I am now completely the victim of our relationship and our success, so I have to develop my own topic independently of you. You should not even touch anything with what I will start from scratch.'" This was the launchpad for the work that Zitvogel is most known for: the microbiome and its effect on cancer. "I decided I will show to the whole world that not only am I autonomous, but I can create from scratch something that doesn't exist at all. And I created the introduction of the importance of the gut microbiota in the responses to any kind of anticancer therapeutics."

> "I decided I will show to the whole world that not only am I autonomous, but I can create from scratch something that doesn't exist at all."

It was nothing short of a revelation.

The microbiome is the diverse community of microorganisms found within a given environmental niche. That niche could be your colon or the root system of a tree. The study of the microbiome is to understand how members of the microbial community interact with each other, as well as with the overall environment.

Germ Theory

Zitvogel had been looking at the effects of chemotherapy on the immune system, and the drug she came to focus on was cyclophosphamide. It is an old drug and the backbone of many a chemotherapy regimen. "A lot of chemotherapy will disappear," suggests Zitvogel, "but the only one which will stay forever is cyclophosphamide because it is doing so many things." Yet, much about the drug's specific mechanisms remains unknown. Zitvogel set out to delineate the drug's effect on the immune system.

The first thing her research team noticed was that the drug caused the immune system to deploy a type of T cell called Th17. "At that time they were almost not known, but the first papers describing them said that they were pro-inflammatory cells involved in autoimmune diseases." It was also known that Th17 cells were associated with the gut. "So, I ask, how come we are getting circulating Th17 cells [outside the gut] with cyclophosphamide?"

Her experience treating patients would provide the next clue. "I am an M.D. by training, an oncologist, and I had been seeing patients for 30 years," often prescribing cyclophosphamide. "I would have these consultations: 'How are you?' And they say, 'Well, my belly this and that, my gut.' The gut, the gut, the gut. It would come back to that over and over again: vomiting, diarrhea, anorexia, lack of appetite. Gut disorders. I said, 'Aha!' So, any drug we are prescribing is hitting the gut." That's how Th17 cells were being activated. The next key question was whether there is a connection between what the drugs were doing at the level of the gut microbiome and the clinical outcome with regard to the immune system.

Some people thought it was a silly question, Guido Kroemer included. "He said that it was completely crazy, totally *phantasmagoire*." Undeterred, Zitvogel pressed ahead with her investigations. "The first experiment was to treat mice with antibiotics and see whether the drug still works." The other approach was to do the experiment with germ-free mice. Results for both approaches were as clear as could be: The cyclophosphamide stopped working. "It took us a while to understand what was going on, but that was the first step toward discovering how important it was." The bugs are helping the drugs.

Checkpoint Please

The experiments described above established a proof of principle. The next step was to see how far the observation could be expanded. "Once we were convinced that the gut microbiome has such a huge influence for cyclophosphamide, we turned to ipilimumab." The most common side effects of ipilimumab are colitis, a serious, potentially lethal problem in the gut, as well as skin rash. At first glance, an association between ipi and the microbiome was readily apparent. "Think about where the side effects occur— like colitis in the gut—this is a portal of entry where the microbes are residing," explains Zitvogel. "And then there's the rash, suggesting skin micro-biota, and then in the liver you see transaminase elevations [a sign of liver function inflammation], and you know that the liver is being drained by the gut … you realize everything had to be explained by the microbiome."

Zitvogel performed roughly the same mouse experiments with ipi that she did with cyclophosphamide, and the results were the same: Without the microbes, ipi was useless. Another research group (Silvan et al., *Science* 350: 1084 [2015]) showed the same results for anti-PD-1 therapy: no microbes, no cure.

Why does this make sense? "There are several mechanisms," Zitvogel suggests. "First, the 'leaky gut' idea, that's quite famous," meaning that under certain conditions, like chemotherapy, the gut actually becomes per-meable to the point that microbes can leak out into the bloodstream. "Another mechanism could be some molecular mimicry between the microbes and the tumor antigens."

To understand this last bit is to realize that our intestinal biota have been with us for a very long time. T cells are accustomed to this; they've been trained. Furthermore, because we've coevolved—us and the bugs—microbial parts often resemble people parts on the molecular level. Although many of the details of this mechanism remain elusive, it is presumed that the microbes are an essential part of educating the immune system to see disease, including cancer.

Bug Drug

Zitvogel might love science, but her endgame is clinical translation: treat-ing patients. The obvious next step was to figure out which microbes were required for immunotherapies to be effective. To do that, an investigator first needs to establish a microbial baseline, the normal microbial population that is typically residing in a healthy gut. That baseline is then compared with the microbial residents of a person who had a suboptimal response to immu-notherapy. Taking advantage of the extraordinary microbiological resources of the Pasteur Institute, Zitvogel discovered that the mice that failed ipi treat-ment lacked a microbe called *Bacteroides fragilis*. When she put the *B. fragilis* back into microbe-deficient mice through means of gavage, or by immuni-zation with fragments of *B. fragilis* via injection, or by adoptive transfer of *B. fragilis*–specific T cells, the activity of ipi was restored. Proof of principle: A 'bug' could be a drug.

Indeed, there's already precedent for microbial therapy. Numerous clini-cal investigations are currently underway to show the efficacy of fecal trans-plants for patients with intractable infections of *Clostridium difficile*, a chronic, debilitating condition that is often the result of both extended hospitalization and the overuse of antibiotics. These "poop transplants" from a healthy donor (frequently, the patient's spouse; read into that what you will) can correct dysbiosis, an imbalance in one's microbial community. In this case, the overgrowth of *C. difficile* and its related pathogenic effects are corrected by a fecal transplant. As Zitvogel considers how to treat dysbiosis in cancer patients, such transplants would be one option. "You could conceive of doing a fecal microbial transplantation of, you know, 'golden shit' from a donor. You could imagine that."

A second approach would be to identify the specific microbe that's missing, and then administer a pill containing that microbe. Refining that concept, one could simply administer the molecules that the missing microbe previously provided, as it's likely that it's not the whole microbe that is affecting the immune system, but rather some component or by-product thereof. A third approach, far more general, would be the use of probiotics. For now, Zitvogel is focused on therapeutics using specific microbes and microbe by-products and is currently in the process of forming a company to support the validation and clinical development of these ideas.

Science, Art, and Life

"You can be an outstanding scientist if you have a lot of, you know, *des grains de sable dans les rouages*," opines Zitvogel, which translates roughly as "sand in the machine." You have to have an idea stuck in your mind, a problem, a vexation, an irritating unknown that you just can't get rid of, that you refuse to let go. "You have to be able to be in the shower, or driving your car, or whatever and only thinking about your experiments and the data, the new data, and how it fits or doesn't fit with the hypothesis." You have to be obsessed, immersed.

Zitvogel cautions, however, against being a robot. "I love painting, even though I am not talented, I admit, but this is something that I like to do. I also like to play the saxophone, and the piano, and I was a dancer when I was young . . . ballet. So, yes, I am in love with all the other things too. I am an artist, I think. This entire scientific thing is artistic. So, you know, creation, and then escape. You escape from the routine of life or the bad things in life." It is a deliberate act, a knowing choice, a strategic retreat to the garden where the seeds of creation are formed, "an escape to a place where you can dream."

Tout commence par un rêve.

Epilogue

Everything in this book is outdated. Not the people, certainly, but much of the science. Even if the writing of this book took just a few weeks to complete, instead of the better part of two years, it would still qualify as old news because the cancer immunotherapy field is moving just that fast. Every month a hundred new IO journal articles are published, every day a thousand new experiments are performed, and it seems like every other week there is another front-page headline.

It's a new-knowledge whirlwind, and even the scientists themselves can't keep up (but it is great fun to watch them attend each other's lectures in the attempt).

On the business front, in the interim between submission and publication, an IO-related biotech company will have boomed, or blown up, and investors—to the tune of hundreds of millions of dollars—will have doubled down or bailed out. By the time this book sees print it is entirely likely that one of the CAR constructs (Section IV) will receive an unprecedented FDA approval for the first-ever such medicine, a therapeutic that is, as one scientist put it, "a living drug."

Conversely, it would be no surprise if one of the technologies highlighted herein has been discounted, and the related company dissolved. This is a high-stakes business with high-risk science. It happens.

What will not happen is that immunotherapy will go away. Every scientist profiled here and every scientist with knowledge of the space will tell you that they've seen nothing like it. Yes, the previously proposed magic cancer bullets have all in some way missed their mark, but this—this discovery and deployment of the arbiters of the immune system (shoot holes in it if you will)—is a rebalancing of the natural forces you were born with. Unlike radiation and chemotherapy, which impose a harsh rule over an errant physiology, immunotherapy simply restores or augments a previously existing order.

It's a no-brainer. It just makes good science sense, and the data bear that out.

From this time forward, when a cancer patient is told for the first time that he or she has cancer, immunotherapy will almost certainly be part of that initial conversation. This is not the future. This is now.

To facilitate the needs of those seeking a greater understanding of the now, there is a website dedicated to that purpose: acurewithin.org.

Over the next year (at least) and continuing as long as the interest warrants the effort, the science detailed in the book will be updated, and new investigator profiles will be posted—both the luminaries not included here and, just as importantly, the next-generation, rising stars.

Therefore, it is openly acknowledged that this book is a beta version, and that updates, patches, and more profiles are required.

The project will continue, because everyone deserves a cure.

Glossary

Activation-induced cytidine deaminase (AICDA): An enzyme that catalyzes the removal of amino groups from the nucleotide base cytosine, converting it to the base uracil.

Adaptive immune response: The arm of the immune system that responds specifically after it encounters a pathogen (as opposed to the innate immune response, which is always primed and ready for general threats).

Allogeneic: Cells or tissues from genetically different individuals of the same species.

Alternative pathway: One of the three complement pathways, it is activated by C3 hydrolysis in the presence of pathogens or foreign materials. Activation does not require formation of an antigen–antibody complex.

Amino acids: The basic building block of proteins. They are characterized by an amine group ($-NH_2-$) and carboxyl group ($-COOH-$), along with a functional group ($-R$). Proteins are formed by peptide bonds between amino acids.

Amyotrophic lateral sclerosis (ALS): A disease affecting the motor nerve cells of the brain and spinal cord that causes a loss of voluntary muscle function over time.

Antibody: A Y-shaped protein that is produced by B cells in response to a specific antigen. These proteins have the ability to bind to harmful substances produced by pathogens.

Antigen: Any external or internal protein that can cause the production of antibodies by the immune system.

Antigen-presenting cell (APC): An immune cell that has the ability to internalize pathogens and display antigens on its surface. The antigens are presented bound to MHC class II molecules.

Arthus reaction: A hypersensitive reaction to intradermal injection of antigens resulting in inflammation and fluid buildup at the site of injection.

Autocrine: A form of cell signaling in which the cell produces hormones or chemical agents that bind to the same cell's receptors and cause an effect in that cell.

Autoimmune: A condition wherein the immune system regards a "self"-antigen as nonself and causes an immune response.

Autologous: Cells or tissues isolated from a person that are re-introduced into the same person after manipulation.

B cells: A type of lymphocyte produced in the bone marrow. These cells have the ability to produce antibodies.

Biomarker: A substance in the body that can be measured to indicate ongoing molecular processes in the body.

CD4: A surface marker present on a subset of T cells known as T helper cells. It acts as a co-receptor for the T-cell receptor by binding to MHC class II molecules on cells.

CD8: A surface marker present on a subset of T cells known as cytotoxic T cells. It acts as a co-receptor for the T-cell receptor by binding to MHC class I molecules on cells.

CD19: A surface marker on B cells that functions as a co-receptor for B-cell antigen receptor to reduce the signaling once the B-cell antigen receptor is bound by antigens.

CD28: A surface marker present on T cells that is needed for T-cell activation. In addition to the T-cell receptor binding MHC class II on target cells, CD28 needs to bind its partner molecule (B7 ligands) on the same target cell.

Cancer immunoediting: The process of the immune system fighting and sculpting tumor growth. It consists of three phases: elimination, wherein the immune system removes most of the cancerous cells; equilibrium, wherein the remaining tumor cells co-exist with the immune system; and escape, in which the tumor cells start forming tumor masses aided by the immune system.

cDNA library: A library of complementary DNA (cDNA) fragments that represent the expressed genome of an organism. cDNA fragments are inserted into a vector and cloned to create a cDNA library.

Chemokine: A messenger protein that belongs to the cytokine family of signaling molecules. It can act on neighboring cells to guide them to the target site.

Chimeric antigen receptor: T-cell receptors that are engineered using viral vectors to be specific for a desired antigen. It is composed of protein fragments from different animal species.

Cloning: Making a copy.

Complementary: A nucleotide on one strand with an appropriate nucleotide base on the other: adenine (A) binds to thymine (T; in DNA) or uracil (U; in RNA) and vice versa; guanine (G) binds to cytosine (C) and vice versa. If the nucleotide sequence of one strand is –AGCTGCTTAC–, then the complementary strand would be –TCGACGAATG–.

CRISPR–Cas9: A gene editing technique that allows researchers to manipulate target genes.

Cross-presentation: When antigen-presenting cells present an antigen bound to MHC class I molecules to CD8$^+$ cytotoxic T cells.

Cross-priming: Stimulation of cytotoxic CD8$^+$ T cells by antigen-presenting cells. The outcome of cross-presentation.

CTLA-4: The inhibitory homolog of CD28. This molecule can bind to B7 ligands with higher affinity and avidity than can CD28, and it inhibits T-cell activation.

Cyclosporin: An antifungal compound produced by fungi that can suppress the immune system.

Cytokine: A family of signaling molecules that have a conserved structure. These molecules are involved in immune cell signaling.

Cytoplasm: The nonnucleated part of a cell's interior, which contains proteins, glucose, and other nutrients, along with organelles.

Cytotoxic T cells: A subset of T cells that are CD8$^+$ (i.e., express CD8) and can kill damaged or infected cells once activated.

Darwin, Charles: An English scientist who proposed the theory of natural selection to explain the change in species over time.

Dendritic cell: An immune cell that can phagocytize pathogens and dead cells. It also functions as an antigen-presenting cell.

Deoxyribonucleic acid (DNA): Deoxyribonucleic acid is the genetic informational code that dictates gene expression and hence all of a cell's characteristics. Each strand of DNA is composed of multiple subunits (nucleotides), each composed of a sugar molecule (deoxyribose), a phosphate molecule, and one of four nucleotide bases (adenine, guanine, cytosine, and thymine).

Expressed sequence tags: Fragments of cDNA clones.

Fab (fragment, antigen binding): The upper, and variable, portion of Y-shaped antibodies. They bind to, and are specific for, specific antigens.

Fc receptor: A receptor present on the surface of some cells capable of binding to the lower portion of Y-shaped antibodies (i.e., the Fc [fragment, crystallizable]).

Gene: A specific set of DNA code that dictates the expression of a specific protein. For gene expression to occur, the information in DNA is transcribed to RNA, which acts as a messenger to translate the genetic code to form a protein.

Graft-versus-host disease (GvHD): A medical complication that can arise when tissue from a genetically different person is transplanted to a new host. The donor's immune cells in the tissue graft recognize the host's cells as nonself and begin attacking them.

Granulocyte-macrophage colony-stimulating factor (GM-CSF): A cytokine released by immune cells that promotes the proliferation and maturation of macrophages and granulocytes.

HeLa cells: An immortalized line of cervical cancer cells isolated from a person named Henrietta Lacks that have been used for many years by researchers to study human cell biology and cancer biology.

Hematopoietic stem cell: A stem cell population found in small numbers in the bone marrow that gives rise to all immune and red blood cells.

Humanized antibody: An antibody isolated from an animal that has been manipulated to contain fragments of human antibodies to reduce immune rejection when administered in humans.

Hybridoma: The technology by which monoclonal antibodies are generated. The method involves fusing a B cell that can produce antibodies against a specific antigen with a cancer cell myeloma to produce a cancerous B cell that will generate antibodies indefinitely.

Immortalized cells: Cells that can grow and multiply indefinitely because of a mutation.

Immune surveillance: A surveillance process used by the immune system to monitor any abnormal cells.

Immunoglobulin: Another name for antibody.

In vitro: Literally, in glass (Latin). A process that takes place outside a living organism (e.g., in a test tube or culture dish).

Induced pluripotent stem cells (iPSCs): Differentiated cells that can be regressed to a pluripotent state by the overexpression of certain genes.

Inflammation: A response of the immune system to a pathogen or foreign intruder. The response is mounted to remove the insult.

Innate immune response: The arm of the immune system that is always prepared for an attack and can respond to general threats. It is not specific or selective to threats (as is the adaptive immune response).

Interferon γ: A damage-signaling cytokine that is released by cells in response to an insult such as a pathogen or virus. This cytokine signals the immune cells to mount an immune response.

Interleukin 2 (IL-2): A cytokine released by T helper cells that is necessary for T-cell proliferation, activation, and maturation.

Islet cell: A cell in the pancreas that is found in the pancreatic islets of Langerhans. These cells produce important hormones that regulate glucose metabolism and growth.

Kinase: An enzyme that can add a phosphate group (from ATP) to another molecule.

Knockout: An organism is which a gene is replaced or disrupted such that it is inactive.

Lamarck, Jean-Baptiste: A French biologist who expanded the field of biology by emphasizing the importance of invertebrates and suggesting that genetic traits are inherited.

Leukapheresis: The process of separating white blood cells from a sample of blood.

Library (phage): A library of proteins that are screened using bacteriophages (i.e., viruses that infect bacteria). A gene of interest is inserted into a virus such that the protein encoded by that gene is displayed on the surface of the virus; the protein thus displayed can then be used as a probe to detect its interactions with other genes and/or proteins.

Ligand: A signaling molecule that can bind to a receptor to initiate intracellular events.

Lupus erythematosus: A disease in which the body's immune cells attack other body cells (e.g., in the skin, joints). A telltale characteristic of this disease is the development of a rash shaped in the form of a butterfly on the person's face.

Lymphocyte: A type of white blood cell. It applies to all cells derived from a common lymphoid progenitor: B cells, T cells, and NK cells. B and T cells are involved in the adaptive immune response, whereas NK cells are part of the innate immune response.

Lymphokine-activated killer (LAK) cells: Natural killer cells that target and kill tumor cells after exposure to the cytokine IL-2.

Macrophage: A type of white blood cell. A common myeloid progenitor gives rise to monocytes, which circulate in the blood before residing and maturing in the tissue to give rise to macrophages. These cells are antigen-presenting cells.

Major histocompatibility complex (MHC): A set of protein complexes involved in presenting antigens to T cells. These protein complexes can be different in individuals and can determine if the tissue from a donor will be compatible with the recipient.

Mast cells: A type of white blood cell derived from a common myeloid progenitor. These cells have granules that can release histamine and play a role in the allergic response.

Meselson–Stahl experiment: An experiment conducted by Matthew Meselson and Franklin Stahl that demonstrated that DNA replication is semi-conservative (i.e., when DNA replicates, each new double strand of DNA is composed of one original strand and one newly synthesized strand).

Metastasis: The process of tumor cells traveling in the body to form secondary tumors in other sites.

Microbiome: The combined genetic material of microorganisms in an environment.

Monoclonal antibody: An antibody that is specific to a target site on the antigen.

Mutant: Refers to the phenotype of an abnormal form of a naturally occurring (i.e., wild-type) species.

Natural killer (NK) cells: A type of white blood cell derived from the common lymphoid progenitor. It can directly bind to pathogens, tumor cells, and virally infected cells and kill them.

Neutralizing antibodies: Antibodies that can bind to proteins and inhibit the binding of those proteins to receptors.

Neutrophil: A type of white blood cell derived from the common myeloid progenitor. It is the first responder to injury and can recruit other immune cells to the site of injury. It is the most common type of white blood cell found in the body.

NF-κB: A transcription factor that is involved in cytokine production and cell survival.

Oncogene: A gene that is associated with a high risk of cancer when overexpressed or mutated in the body.

Paracrine: A form of cell signaling in which the cell produces chemical signals that bind to a nearby cell's receptors and causes an effect in that cell.

Phagocytosis: The process by which a cell engulfs/ingests a pathogen or abnormal cell and breaks it down within the cell to inactivate it.

Pharmacokinetics: The branch of pharmacology that examines how the body reacts to the administration of drugs.

Phosphorylation: The process of adding a phosphate group to a molecule (usually protein).

Pipette: An instrument used to measure out precisely small volumes of liquids, usually in the 0.1- to 1000-microliter range.

Plasma: The fluid component of blood in which cells are suspended. It accounts for 55% of blood's total volume and contains dissolved substances such as oxygen, glucose, and proteins.

Protein: A macromolecule composed of amino acids linked together to form complex structures. These are the end products of the instructions of genes.

Receptor: A protein on the surface of a cell or within its nucleus that can bind a ligand to cause signaling events in the cell.

Recombinant (DNA, protein): Not naturally occurring. Usually refers to molecules that are engineered outside the body.

Response Evaluation Criteria In Solid Tumors (RECIST) criteria: A set of rules to characterize how tumors progress during clinical treatment.

Retrovirus: Viruses that have RNA as their genetic material. These viruses contain the enzyme reverse transcriptase that can synthesize a single cDNA strand from the RNA. That single strand of viral cDNA is then converted via the infected cell's machinery to a double-stranded sequence, and expression of the viral gene then proceeds by the normal cellular mechanisms.

Ribonucleic acid (RNA): RNAs are nucleotides composed of a sugar molecule (ribose), a phosphate molecule, and one of four nucleotide bases (adenine, guanine, cytosine and uracil [as opposed to thymine, as in DNA]). For gene expression to occur, the information in DNA (in the cell's nucleus) is transcribed to a complementary sequence of RNA, which transports the code to the cytoplasm where it can be used to instruct the cell to form a protein.

Sequence (DNA; amino acid): A particular order of the building blocks of DNA (nucleotide bases) or proteins (amino acids).

Stem cell: An undifferentiated cell that can produce copies of itself or differentiate into a defined cell type.

Subtractive hybridization: A technique to identify and characterize differences between nucleic acid sequences in cells from different tissues or different growth phases or after treatment with a drug.

T cells : A type of lymphocyte.

T helper cells: A subset of T cells that are CD4$^+$ (i.e., express CD4) and are necessary for activation of cytotoxic T cells.

T-cell receptor (TCR): A receptor present on T cells needed for T-cell activation. The TCR on T cells binds to MHC class I/II molecules on other cells and initiates signaling events in the T cells.

Th17 cells: A subset of T helper cells characterized by their production of the cytokine interleukin-17. These cells are involved in the adaptive immune response.

Thymic selection: A selection process that takes place in the thymus eliminating strongly self-reacting T cells.

Toll-like receptors (TLRs): A class of proteins present on immune cells that can recognize structurally conserved molecules released by pathogens. They play a key role in the activation of the innate immune response.

Transcription factor: A type of protein present in the nucleus that can promote or repress transcriptional activity of target proteins.

Translation (clinical): Taking a process or technology from the laboratory bench to the bedside.

Tumor microenvironment: The cellular environment surrounding a tumor, which contains many cell types, stroma, immune cells, and blood vessels.

Tumor-infiltrating lymphocytes (TILs): T cells that have migrated from the blood circulation into the tumor tissue.

Type 1 diabetes: A metabolic disease in which the body cannot produce sufficient insulin to regulate glucose levels in the body.

Type I interferon: Proteins released by leukocytes in response to viral infection. This protein enhances the activation of NK cells and macrophages to kill the infected cells.

Vector (viral): The DNA backbone of a virus that acts as a vehicle to deliver foreign genetic material into cells, after the harmful viral parts have been removed.

V(D)J recombination: Variable (V), diversity (D), and joining (J) genes can assemble themselves randomly (i.e., recombination) to generate a diverse set of antigen receptors. This recombination process allows B cells and T cells to recognize and battle an indefinite number of antigens.

Index